CONTEMPORARY'S

Critical Thinking With Math

Reasoning and Problem Solving

Karen Scott Digilio

Senior Editor
Mark Boone

 **Wright Group**

Wright Group

ISBN: 0-8092-4455-1

Send all inquiries to:
Wright Group/McGraw-Hill
130 East Randolph Street, Suite 400
Chicago, Illinois 60601

Printed in the United States of America.

4 5 6 7 8 9 GB(H) 23 22 21 20

The McGraw-Hill Companies

Contents

To the Instructor

Critical Thinking with Math is designed to help students develop the reasoning and problem-solving skills needed to succeed in today's world. Addition, subtraction, multiplication and division skills are not enough by themselves; the ability to recognize a problem and reason it through to a solution is vitally important.

This book is divided into two sections—**problem solving with whole numbers** and **decimals, fractions and percents**. In each section, thinking skills move from the simple to the more complex as lessons gradually build on students' previously acquired competencies. For this reason, it is important that the book be completed in the order in which the skills are presented.

Critical Thinking with Math exposes students to number relationships, word-to-number translations, problem-solving strategies, practical applications of math, and techniques for simplifying computational procedures. These are presented in the clear format of a short introduction, an example with explanation, and ample practice exercises.

Critical-thinking skills are often correlated to the hierarchy of Bloom's taxonomy. The math skills included in this book may be classified under the cognitive levels of comprehension, application, analysis, synthesis, and evaluation. The following list correlates some of the math skills in this book to these levels:

Cognitive Level	Skill
comprehension	performing the four basic arithmetic operations understanding what a problem asks for
application	using the four basic arithmetic operations to solve simple word problems
analysis	identifying situations in which either not enough or too much information is given
synthesis	putting several skills together to solve multiple-operation problems
evaluation	comparing and contrasting patterns among word problems

The purpose of this book is to strengthen thinking skills and problem-solving abilities. This differs from the traditional math text which focuses on computations and their applications to word problems. For this reason, you should make sure that students have mastery of the required mathematical skills before they attempt the problem-solving material. Otherwise, they may be frustrated in their attempts to master a reasoning concept.

This book contains a detailed answer key at the end that will help students understand their mistakes. The answer key is designed to allow a student to use the book independently. Generally, however, this type of book and the levels of skills it teaches are best presented under the guidance of a teacher. Therefore, it is highly recommended that the instructor work through the material with the students section by section.

This worktext can be used most effectively as a springboard for further activity in critical thinking with math. By using some of the instructional approaches included in it, materials can be adapted from newspapers, magazines, and situations in daily life. Whether in role-playing with money, measuring items, or applying their skills to sales advertisements, students should be encouraged to continually practice their reasoning and problem-solving skills in math.

To the Student

Welcome to Contemporary's *Critical Thinking with Math*. In this book, you will learn skills that will help you to solve problems in everyday life.

This book includes the topics of number patterns, estimating and rounding, word-to-number translations, formulas, making diagrams, and problem-solving strategies. All are related to whole numbers, measurement, decimals, fractions, and percents. The skills in this book move from the simple to the more difficult. Therefore, it is best to complete the lessons in the order in which they are presented.

This book has a number of features that will make it easier for you to learn. They include:

- short introductions to each lesson, followed by detailed examples and plenty of practice
- skill levels that gradually get more difficult as they relate new skills to the earlier ones
- practical applications of math skills to everyday life
- an answer key that explains the correct answers. If you make a mistake, you can learn from it by reading the explanation that follows.

Keep in mind, as you complete the exercises in this book, that critical thinking can become a part of your everyday life when you work with math. A good way to continue sharpening the skills you learn in this book is by applying them to practical tasks such as:

- balancing a checkbook
- estimating the cost of multiple-item purchases
- measuring ingredients when cooking
- measuring materials when sewing or doing carpentry
- figuring the discount on sales items

You too can gain the confidence that it takes to be a good problem solver.

1 Problem Solving with Whole Numbers

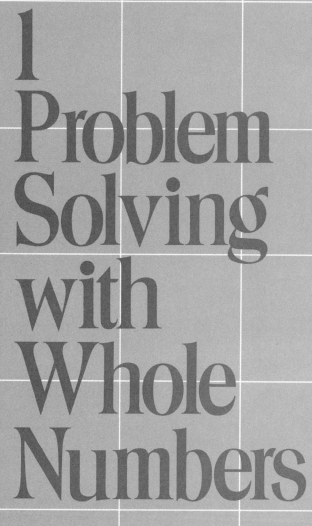

No one is comfortable with uncertainty. Problems bring uncertainty into our lives until they are solved. We all are faced with problems on a daily basis. But the ability to solve them is a skill that must be developed in most of us; most of us are not born with this ability. We must learn formulas, be able to see relationships and patterns, and understand the steps needed to solve a problem. This section will develop your problem-solving skills and prepare you to handle more difficult questions later.

Number Relationships

Think of something that you do well—cooking, sewing, typing, fixing cars, or playing a sport. You probably understand that activity very well and feel good about yourself when you are doing it. An ability to solve problems requires confidence. You can feel comfortable using your math skills only when you completely understand numbers and their relationships.

In this chapter you will work with skills that you will use again and again throughout this book. They include:

- using multiplication facts
- finding patterns and series
- rounding and estimating

The Multiplication Grid

Knowing the *multiples* of the numbers 1 through 10 is important to your success with math in school and in daily living.

▶ Complete the grid below by writing the answer to each pair of numbers multiplied across and down. The example below shows that the result of multiplying 3 × 2 is 6.

	1	2	3	4	5	6	7	8	9	10
1										
2			6							
3										
4										
5										
6										
7										
8										
9										
10										

▶ You can work with numbers more quickly and easily if you see patterns in them. Let's try to find some in the multiplication grid.

1. The second row down (or column across) should list all the multiples of 2—what are called **even numbers**. All even numbers end in _____, _____, _____, _____, or _____.

2. The third row down should list all the multiples of 3. Look at the multiples of 3 starting with 3×4. At this point they become two-digit answers. Multiply the numbers, and then add up the digits.

$3 \times 4 = $ _12_   _1_ + _2_ = _3_

   a. $3 \times 5 = $ _____   _____ + _____ = _____

   b. $3 \times 6 = $ _____   _____ + _____ = _____

   c. $3 \times 7 = $ _____   _____ + _____ = _____

   d. $3 \times 8 = $ _____   _____ + _____ = _____

   e. $3 \times 9 = $ _____   _____ + _____ = _____

   f. $3 \times 10 = $ _____   _____ + _____ = _____

   g. In each case, the digits add up to _____, _____, or _____. These are all multiples of 3.

3. Look at the multiples of 5 now. They are easy to remember.

   All multiples of 5 end in either _____ or _____.

4. **a.** Now study the multiples of 6.

   Are they all even or odd ? _____

   b. Add the digits of the multiples of 6.

   They are also multiples of _____.

5. **a.** Now find the multiples of 9. Add up the digits of the answers.

   They all add up to _____.

   b. Write the multiples of 9 starting with $9 \times 2 = 18$.

Write the Beginning Digits	Write the Ending Digits
<u>1</u>	<u>8</u>
_____	_____
_____	_____
_____	_____
_____	_____
_____	_____
_____	_____
_____	_____

   c. In each case the first digit goes up by _____, and the second digit goes down by _____.

6. Find the multiples of 10. They are easy to remember because they all end in zero. In other words, to find a multiple of 10, simply add a _____ to the number being multiplied.

Knowing multiples can be helpful in making quick decisions about dividing. Remember that:

- numbers ending in 2, 4, 6, 8, and 0 are divisible by: 2
- numbers ending in 0 and 5 are divisible by: 5
- numbers ending in 0 are divisible by: 10

▶ Use your understanding of multiples to answer these questions.

7. Circle the number(s) that are evenly divisible by 2.

   7; 40; 766; 1,423

8. Circle the number(s) that are evenly divisible by 5.

   6; 55; 400; 1,577

9. Circle the number(s) that are evenly divisible by 10.

   1; 70; 103; 1,020

10. Circle the number(s) that are evenly divisible by 5 and 10.

   5; 90; 705

Answers are on page 106.

Prime Numbers

A **prime number** is a number larger than 1 that can be evenly divided only by itself and 1. The number 2 is the first prime; it can be evenly divided only by 2 and 1.

1̸	②	3	4̸	5	6	7	8	9	10
11	12	13	14	15	16	17	18	19	20
21	22	23	24	25	26	27	28	29	30
31	32	33	34	35	36	37	38	39	40
41	42	43	44	45	46	47	48	49	50
51	52	53	54	55	56	57	58	59	60
61	62	63	64	65	66	67	68	69	70
71	72	73	74	75	76	77	78	79	80
81	82	83	84	85	86	87	88	89	90
91	92	93	94	95	96	97	98	99	100

▶ Using the chart above, follow the steps to find the prime numbers under 100. You may cross out a number more than once.

1. Cross out 1.
   (*It is not considered prime.*)

2. Circle 2 because it is divisible only by it self and 1. Cross out all multiples of 2. These numbers are not prime because they are divisible by 2.
   (*These are all even numbers.*)

3. Circle 3 and cross out all multiples of 3. You know that a multiple of 3 cannot be a prime number because it is divisible by itself and 3.
   (*Add the digits of two-digit numbers as a fast check for these; they should all be multiples of 3.*)

4. Circle 5 and cross out all multiples of 5. You know that a multiple of 5 cannot be a prime number because it is divisible by itself and 5.
   (To find these quickly, look at the ending digits 5 and 0.)

5. Circle 7 and cross out all multiples of 7. A multiple of 7 cannot be a prime number because it is divisible by itself and 7.
   (These are hard, but use your multiplication grid to help you get started.)

6. Circle 11; then cross out all multiples of 11. A multiple of 11 cannot be a prime number since it is divisible by itself and 11.
   (HINT: The multiples of 11 follow a pattern. All under 100 are twin digits, like 22.)

7. Circle 13 and cross out all multiples of 13. A multiple of 13 cannot be a prime number since it is divisible by itself and 13.
   (These are difficult. You may need a scratch pad to figure these!)

8. Go to the next number that is not crossed out, circle it, and find its multiples. You may need scratch paper to do many of these. Continue the pattern until all numbers under 100 have been either circled or crossed out.

9. Now list in order all prime numbers under 100. _____

   _____

10. How many did you find? _____

Answers are on page 106.

Finding the Pattern

The more you work with math, the more you will see the need to identify *patterns* that develop out of number relationships. You discovered some rules or patterns when you worked with the multiplication grid and prime numbers.

Let's look at an easy example. Fill in the boxes.

2 ■ 3 = 6   8 ■ 1 = 8

5 ■ 7 = 35   14 ■ 2 = 28

If you filled in the boxes with the **multiplication sign** ⊠ **(times),** then you were able to see the pattern. The pattern is **to multiply the numbers.**

▶ Fill in the boxes below. Then finish the rule for the pattern.

1. 4 ■ 4 = 8

3 ■ 16 = 19

0 ■ 10 = 10

The pattern is _____

_____.

2. 20 ■ 4 = 5

9 ■ 3 = 3

14 ■ 7 = 2

The pattern is _____

_____.

Let's try something a little harder. Fill in the boxes, and write the pattern.

2 × ■ = 6   6 × ■ = 18

3 × ■ = 9   7 × ■ = 21

The pattern is _____

_____.

In each case, the statement is made true by writing the number 3 in the blank, so the rule is **multiply by 3.**

▶ Write the numbers in the boxes, and then finish the pattern.

3. 4 + ■ = 12

7 + ■ = 15

1 + ■ = 9

The pattern is _____

_____.

4. 3 − ■ = 2

6 − ■ = 5

10 − ■ = 9

The pattern is _____

_____.

Answers are on page 106.

Patterns and Series

Being able to see a pattern in a situation or story is a skill that can help you solve many different kinds of problems.

Can you see a pattern in this series?

Draw the next figure in the series.

You're correct if you drew a triangle with its peak pointing up.

▶ What would be the next letter or letters in these series?

1. A, E, I, O, _____

2. B, C, D, E, F, G, _____

3. R, S, S, T, T, T, U, U, U, U, V, V, V, V, _____

4. A, C, E, G, I, K, _____

5. A, B, Z, Y, C, D, X, W, _____   _____

Words can form a pattern, too. What pattern do you see here?

       big, bigger, _____

The word *biggest* completes the series because the pattern is to list higher levels of the adjective *big*.

▶ Now finish these patterns.

6. good, better, _____

7. ready, set, _____

8. man, woman, boy, girl, male, _____

9. forward, back, back, forward, back, back, _____

10. September, April, June, _____
(HINT: Think about what these months have in common.)

Answers are on page 106.

Measurement and Patterns

One use of patterns is to show the relationship between different types of measurements. Look at the series below.

sixteenths, eighths, quarters, _____, wholes

What is missing? _____

What measurement is being shown? _____

Is this an increasing or decreasing relationship? _____

You should have filled in the missing word **halves.** The measurement shown is **fractions.** These fractions are **increasing.** Look at the picture below.

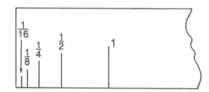

On a ruler, an eighth is bigger than a sixteenth, a quarter is bigger than an eighth, and a whole is bigger than a half.

▶ Fill in each blank with the missing word that completes the pattern. Then tell what type of measurement is being shown and whether the pattern indicates an increasing or decreasing relationship.

1. week, day, hour, minute, _____

   What measurement is being shown? _____

   Increasing or decreasing relationship? _____

2. dollar, half-dollar, quarter, _____, nickel

   What measurement is being shown? _____

   Increasing or decreasing relationship? _____

3. inch, foot, yard, _____

   What measurement is being shown? _____

   Increasing or decreasing relationship? _____

4. _____, quart, pint, cup, tablespoon

   What measurement is being shown? _____

   Increasing or decreasing relationship? _____

Answers are on page 106.

Numbers in a Series

In the last lesson, you found patterns in words, pictures, and measurements. Patterns can also be found in a series.

> For example, 2, 4, 6, 8, 10 . . . is a series of increasing even numbers.
>
> 1, 2, 4, 8, 16, 32, 64 . . . follows the pattern of doubling the previous number.
>
> What is the pattern of this series of numbers? 100, 88, 76, 64, 52 . . .

You probably discovered that each number is **reduced by 12**.

▶ Identify each pattern, and complete the series below. On the line provided, write what the pattern is.

1. 5, 10, 15, 20, 25, _____   The pattern is _____.

2. 118, 109, 100, 91, 82, _____   The pattern is _____.

3. 10,000; 1,000; 100; 10; _____   The pattern is _____.

4. 2, 4, 16, 256, _____   The pattern is _____.

More than one operation can be used in a series. Let's look at an example.

$$1, 3, 2, 4, 3, 5, 4, 6, 5, . . .$$

You probably saw that the pattern shown is to add 2, then subtract 1. You might have arrived at this pattern by writing out the operation like this:

```
1   3   2   4   3   5   4   6   5
 \ / \ / \ / \ / \ / \ / \ / \ /
 +2  -1  +2  -1  +2  -1  +2  -1
```

The pattern is + 2, − 1.

▶ In the number series below, find each pattern and complete it. On the line, write what the pattern is.

5. 1, 10, 8, 80, 78, 780, _____   The pattern is _____.

6. 0, 1, 3, 6, 10, 15, 21, _____   The pattern is _____.

7. 100, 50, 52, 26, 28, 14, _____   The pattern is _____.

8. 606, 600, 612, 606, 618, _____   The pattern is _____.

Answers are on pages 106–107.

Number Patterns

Sometimes number patterns are shown in chart form. Often a few numbers are given, and you have to complete the pattern.

Look at the chart below. What is the pattern? _____

+	3
2	5
4	7
6	
8	
10	

The pattern shown is to **add 3** to the number in the left-hand column in order to get the number in the right-hand column. For example, 3 added to 2 equals 5. Complete the chart by filling in the correct numbers.

▶ Now complete the patterns in the following charts. Then fill in each blank with the word that describes the pattern.

1.

×	5
1	5
3	15
6	
7	
10	

The pattern is to _____ a given number by 5.

2.

−	0
6	6
11	11
26	
30	
108	

The pattern is to _____ 0 from a given number.

Let's look at a pattern that is more difficult:

x	2	3	4	5	6	7	8	9	10
y	20	30							

In this pattern, **x increases by 1 as y increases by 10**. This pattern holds throughout. Following this rule, complete the rest of the chart.

Complete the charts below, and fill in the blank that makes the statement true for each.

3.

x	1	12	20	33	48	51	53
y	7	18	26				

The pattern is x _____ = y.

4.

x	9	18	27	36	45	54	63
y	1	2	3				

The pattern is x _____ = y.

Answers are on page 107.

Estimating

Estimating numbers is helpful in everyday living as well as in tests and books. It can save you time and give you more confidence in coming up with a solution. One way that you can estimate is by rounding numbers.

When grocery shopping, round to the nearest $10.

To estimate the number of people in the United States round to the nearest million.

▶ For each of these situations, circle the nearest value to which it would be best to round.

	Nearest	**Nearest**	**Nearest**
1. buying a wedding dress	$1	$10	$100
2. counting aspirin left in a bottle	10	100	500
3. budgeting monthly rental expense	$10	$100	$1,000
4. guessing the number of spectators at a pro football game	100	1,000	10,000
5. estimating popular votes for a presidential candidate	1,000	10,000	100,000

Answers are on page 107.

Rounding Numbers

1 , 234 , 567

To estimate, you should round numbers. To do this, you will need to understand *place value*. Look at the illustration.

To round a number, look at the digit to the right of the place you wish to round to. Compare rounding $146 and $156 to the nearest hundred dollars:

┌───── hundred
$146
└───── less than 5, so $146 rounds to $100

┌───── hundred
$156
└───── 5 or more, so $156 rounds to $200

Look at the place being rounded to. If the number to the right is less than 5, the number being rounded stays the same. If the number is 5 or more, the place value goes up by 1.

In other words, $146 is closer to **$100**, and $156 is closer to **$200**. All digits to the right of the place being rounded to will become **zero**.

▶ Round these numbers to the nearest value listed.

1. 248 pages—to the nearest 10 pages _____

2. 6,666 people—to the nearest 100 people _____

3. 74,499 boxes—to the nearest 1,000 boxes _____

4. 230,185 births—to the nearest 100,000 births _____

5. $9,520,150—to the nearest $100,000 _____

▶ Compare each pair of problems. Which would be easier to solve? Put a check mark over each one that is easier.

6. a.	585 + 621	b.	600 + 600	8. a.	742 × 356	b.	700 × 400
7. a.	1,782 − 819	b.	1,800 − 800	9. a.	119)‾2,391	b.	120)‾2,400

▶ Now solve problems 6a and b, and 7a and b.

Was it easier to work with the numbers ending in zero? You were able to work the rounded problems in your head by simply bringing down the rows of zeros.

► Now solve problems 8a and b. The rounded one is much easier if you

- bring down all zeros
- multiply the remaining digits

$$\begin{array}{r} 7\,|\,00 \\ \times\ 4\,|\,00 \\ \hline 28\,|\,0,0\,|\,00 \end{array}$$

► Solve problems 9a and b. The rounded one is especially easy if you

- cross out the same number of zeros in each number
- divide the new problem

$$\begin{array}{r} 20 \\ 12\cancel{0}\,)\,\overline{240\cancel{0}} \end{array}$$

Now compare your pairs of answers in problems 6–9. Are they close in value?

► Round the problems below to the nearest ten (for two-digit numbers) or hundred (for three-digit numbers). *Then solve the rounded problems only.*

Actual	Rounded	Actual	Rounded

10. 488
     713
   + 49

12. $316\,)\,\overline{872}$

11. 26,733
     − 9,542

13.   317
     × 854

► For the following problems, round the problems in your head. Then test yourself to see how quickly you can pick the correct estimated answer.

14. $4,826 × 68 accounts =

   a. $35,000   **b.** $350,000   **c.** $3,500,000

15. 5,787 sandwiches ÷ 326 people =

   a. 2 sandwiches each   **b.** 20 each   **c.** 200 each

16. 76 math books + 28 science books + 19 language books =

   a. 100 books   **b.** 130 books   **c.** 150 books

Answers are on page 107.

2 From Words to Numbers

"A picture is worth a thousand words," the saying goes. In math, pictures or symbols are often used to express ideas that would be awkward using words. For example, we use *3* instead of the word *three* and the sign = instead of the words *is equal to*. Your ability to move from words to numbers, symbols, and pictures will help you to solve many kinds of math problems.

In this chapter, you will learn skills that will help you translate from words to mathematical expressions. They include:

- writing expressions and equations
- choosing the operation symbol
- using formulas

Writing Expressions

Many people look for methods to help them solve a problem. One method is an **arithmetic expression**. These expressions contain key words that can be aids in setting up some problems.

Key Words	Meaning	Math Symbol
is	equals	=
of	times (multiplied by)	×
for, per	divided by	÷

Let's look at examples of these key words used in sentences.

3 of the 10-pound packages equal 30 pounds.
(×)   (=)

9 problems per each 3-minute period is 3 problems per minute.
(÷)   (=)

HINT: Notice that *per* is a key word to indicate that division should take place to get the final answer. It is usually found as part of a question, such as "How many miles per hour did Jack travel?"

▶ For each expression, circle the key words shown in the table and write the math symbols below them.

1. 1 dog + 2 cats + 2 birds + 5 goldfish is 10 pets.

2. 500 pages for 2 weeks is 250 pages per week.

3. $48 for 2 shirts is $24 per shirt.

4. 3 copies of each of 36 photographs is 108 photographs altogether.

5. 8 dresses of each of the 3 styles is 24 dresses altogether.

The next step is to write an **expression** (sometimes called a **number sentence**) using the numbers and symbols for the words. Here is an example:

If Mason paid 60 cents   for 12 pencils, how much did he pay per pencil?

60 cents ÷ 12

▶ Write an expression for each situation below. *Do not solve the problems.*

6. How much did Sarah pay per tire if the cost was $160 for four tires?

7. What was the total weight of the men if Brian saw 7 of the 200-pound men?

8. How many crayons were there if Rosa had 10 boxes of the 24-crayon size?

9. What was the size per class if there were 28 students for 2 teachers?

Answers are on page 107.

Equations and Unknowns

Sometimes expressions use letters to stand for missing numbers. The missing number is called an ***unknown***. Unknowns are often used in equations. An ***equation*** includes an equal sign and a letter or symbol for the unknown number. In an equation, the expressions on each side of an equal sign must be equal. Here is an example:

How much cola is in 8 bottles of 16 ounces each? In words, an ***expression*** would be: 8 bottles of 16 ounces each.

The amount of cola in 8 bottles is the unknown. Let a stand for the unknown. The equation would be set up like this:

$$8 \text{ bottles} \times 16 \text{ ounces} = a$$

To solve the problem, you multiply 8 by 16.

The solved equation would read:
8 bottles $\times$ 16 ounces = 128 ounces, so
$$a = 128 \text{ ounces}$$

HINT: Any letter can be used to stand for an unknown. Most often, x is used.

▶ For these problems, write an expression and an equation, and then solve.

1. How much does one kiwi fruit cost if the price is set at 50 cents for 2?

EXPRESSION: _____

EQUATION: _____

SOLUTION:

2. If Gloria buys 6 of the $5 hand towels for gifts, how much will she spend for all?

EXPRESSION: _____

EQUATION: _____

SOLUTION:

3. How much wattage is consumed by 11 light bulbs of 60 watts each?

EXPRESSION: _____

EQUATION: _____

SOLUTION:

▶ Let's write and solve equations for the next problems. They look more complicated, but they are basically the same as those you have just finished.

4. A company offers a $324 monthly bonus for each 3-person workstation if no one at the workstation misses any work. How much will the bonus be per person?

EXPRESSION: _____

EQUATION: _____

SOLUTION:

5. Twelve 52-pound bales of hay are loaded into a half-ton truck. How much weight is being carried in the truck?

EXPRESSION: _____

EQUATION: _____

SOLUTION:

Answers are on page 107.

Choosing the Operation

On pages 16 and 17 you learned that certain key words translate to mathematical symbols (*is* means =). In other cases, key words can tell you whether to add, subtract, multiply, or divide. However, key words can sometimes be confusing.

For example, the word *total* often indicates a multiplication or addition problem. This is because you find a total when you combine numbers. But look at this problem. Is it an addition or subtraction problem?

> A total of 25,126 people attended a pro football game, but 3,963 of them had free guest passes. How many paid to get in?

In this problem, are you being asked to find a total? No, you use the total to find a *difference* or a *leftover* amount. When you find a difference or leftover amount, the problem is a **subtraction** problem.

In this lesson, you will learn how to identify which operations to use without just relying on key words. **Addition** and **multiplication** are generally used when you are asked to find:

- a total
- a combined amount
- a larger figure
- a multiple of something

Look at the problem below. Do you add or multiply?

> Allison was receiving a salary of $325 a week before getting a $25 weekly raise. Now what does she earn each week?

This is an *addition* problem because the situation asks you to find _____

_____

You're right if you said something like **Allison's new salary**.

Now look at the problem below. Do you add or multiply?

> Arthur started out at his job earning $3 an hour. He now earns three times that amount. How much is his hourly wage now?

This is a *multiplication* problem because the situation asks you to find

_____.

You're right if you said something like **three times his hourly wage**.

▶ Put the symbol + by each problem that is an addition problem and a × by each problem that is a multiplication problem. If the problem is neither addition nor multiplication, leave the line blank. Do not solve the problem.

_____ 1. A women's club collected a total of $1,535 one year. They decided to split it evenly among 5 charities. How much money did each charity receive?

_____ 2. Lucy has been attending Weight Watchers for six months. She lost 5 pounds the first month, 3 pounds the next, then 4 pounds, 2 pounds, 2 pounds, and finally 3 pounds. How much did she lose altogether?

_____ 3. A basketball player scored 20 points each game for 7 games in a row. How many points did he score in 7 games?

_____ 4. The temperature was 81 degrees at 8:00 one morning and went up 15 degrees more by 2:00 in the afternoon. What was the temperature at 2:00 P.M. that day?

_____ 5. The Parillas spent $10,455 on their new car. This was $2,015 more than they had spent on their first car. How much was their first car?

Now that you've seen problems that require you to add or multiply, look at those involving subtraction or division. **Subtraction** is generally used to find:

- a difference
- a reduction
- a leftover amount

Rafael earned $16,871 last year and $19,428 this year. How much more did he earn this year than last year?

This is a *subtraction* problem because the situation asks you to find

_____.

You're right if you said a **difference**.

Division is generally used to find:

- equal pieces or amounts of something
- an average of several figures

Three friends went out to dinner together. Their bill totaled $48. The friends decided to share the bill evenly. How much did each pay?

This is a *division* problem because the situation asks you to find

_____.

The correct answer is **equal pieces**.

▶ Put the symbol − by each problem that is a subtraction problem and a ÷ by each problem that is a division problem. If the problem is neither subtraction nor division, leave the line blank.

_____ **6.** In a certain office building, there are 40 workers on each of nine floors. How many people work in the building?

_____ **7.** A clerk has 105 documents to file. If he can file 15 every hour, how long will it take him to complete the task?

_____ **8.** If a dozen eggs costs 96 cents, how much does each egg cost?

_____ **9.** If Kim reads 42 pages a night, how long will it take her to finish a 294-page book?

_____ **10.** The high temperature for the day was 71 degrees, and the low was 39 degrees. What was the temperature change that day?

Answers are on page 107.

The Averaging Formula

Some types of problems are so common that you can use a formula to solve them. A **formula** is a statement in words or symbols that you can always use to solve a certain type of problem. Of course, first you have to identify the type of problem.

One type of problem you can solve by using a formula is finding the average of a set of numbers. It is an easy problem to recognize because it usually contains several figures and the word *average*. Let's look at an example:

The average high temperatures in Washington, D.C., one summer were June—80 degrees, July—85 degrees, August—84 degrees, and September—75 degrees. *What was the average high temperature for the whole summer in Washington?*

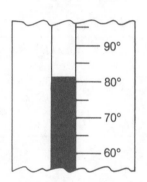

June	80°
July	85°
August	84°
September	75°
Total	324°

$$\begin{array}{r} 81° \leftarrow \text{average} \\ 4\overline{)324°} \end{array}$$
number of months ╱   ╲ total

Therefore, the average high total is **81 degrees**.

▶ Use another sheet of paper to find the averages in each of these problems.

1. Hugh made scores of 79, 82, 80, 84, and 75 on his German tests. What was his average score?

2. At a small evening school one night, there were 15 people in a woodworking class, 21 people in a Spanish class, 23 people in a GED class, 27 in a real estate class, 19 in a computer class, and 15 in an exercise class. What was the average attendance per class that night?

3. Fran spent $105 one week for food for her family, $92 the next week, $110 the next, and $101 the last week of the month. On the average, what did Fran spend weekly on food that month?

4. A business manager logged 4,050 miles in flight travel one month, 2,106 miles the next, and 2,922 miles the last month of the quarter. How many air miles did he average monthly that quarter?

5. In a doctor's waiting room one day were seven expectant mothers aged 17, 23, 24, 26, 32, 34, and 40. What was the average age of expectant mothers in the doctor's office that day?

Answers are on pages 107–108.

The Distance Formula

You just learned how one formula helps you find averages. You can also use a formula with **distance problems**. Let's look at an example:

> On a trip, the Gomez family found they could average 45 miles per hour (mph) for 6 hours of daily driving. How far will they travel in a day?

Notice that the question asks "how far?" These words, or something similar, will be used in most distance problems. Other words that indicate a distance problem are "miles per hour" and "hours." The distance formula is $D = RT$.

Distance (D)	=	Rate (R)	×	Time (T)
Usually expressed in miles		Speed: usually expressed in miles per hour (mph)		Usually expressed in hours or parts of an hour

In the problem above, indicate the Gomez family's speed by underlining it and writing an R above it. Underline the time and write a T above it. Then find the distance by using the formula $D = RT$.

You should have circled **45 mph** and labeled it as the rate; **6 hours** and labeled it as the time; and you should have found the distance to be **270 miles**: $D = 45 \times 6 = 270$.

▶ Solve these problems by identifying the rate (R) and time (T), and multiplying them together to find the distance (D). Notice that distance is the unknown in these problems.

1. How much distance could a plane cover by averaging 350 miles per hour for 4 hours?

$$D \quad = \quad R \quad \times \quad T$$

_____ = _____ × _____

2. A wrecked car is being towed at the rate of 35 mph. If it takes 3 hours to get to the junkyard, how far away is it?

$$D \quad = \quad R \quad \times \quad T$$

_____ = _____ × _____

The same distance formula can be used to find either the rate or time, as well as the distance. Look at this example:

The Johnsons average 40 mph on their trip. They have 360 miles to travel. How long will it take them to get there?

$$D = R \times T$$
$$360 \text{ miles} = 40 \text{ mph} \times T$$
$$360 \div 40 = T$$
$$9 = T$$

In this problem, the distance and the rate are already given. To find the time, you must divide the distance by the rate. This is because division is the opposite (inverse) operation of multiplication.

▶ Solve these problems by identifying the rate (R) and the distance (D), and then dividing for the time (T).

3. A truck traveling 40 mph will reach a town 160 miles away in how many hours?

$$D = R \times T \qquad T = D \div R$$
$$\underline{\hspace{1cm}} = \underline{\hspace{1cm}} \times T \qquad T = \underline{\hspace{1cm}} \div \underline{\hspace{1cm}}$$

4. If a towed car needs to reach a point 420 miles away, how long will it take to get there if it is being transported at a rate of 35 mph?

$$D = R \times T \qquad T = D \div R$$
$$\underline{\hspace{1cm}} = \underline{\hspace{1cm}} \times T \qquad T = \underline{\hspace{1cm}} \div \underline{\hspace{1cm}}$$

The following problems are missing the rate; rate is the unknown. Solve in the same way as you solved the time, except that now you will divide distance by the time.

5. If the Wintczaks need to cover 1,100 miles in 20 hours, what speed do they need to average?

$$D = R \times T \qquad R = D \div T$$
$$\underline{\hspace{1cm}} = R \times \underline{\hspace{1cm}} \qquad R = \underline{\hspace{1cm}} \div \underline{\hspace{1cm}}$$

6. A new jet covers 1,953 miles in 3 hours. How fast does it go?

$$D = R \times T \qquad R = D \div T$$
$$D = R \times \underline{\hspace{1cm}} \qquad R = \underline{\hspace{1cm}} \div \underline{\hspace{1cm}}$$

Answers are on page 108.

3 Problem-Solving Strategies

Now get ready to make a big leap forward in your problem-solving skills. An effective plan can help you win any game—including problem solving—with confidence in your ability.

In this chapter you will use strategies that you can use in all types of problem-solving situations. These include:

- restating a problem
- figuring out the question
- using information
- choosing the operation
- making sense out of answers

Restating the Problem

To *restate* something means to put a story or an idea into your own words. This is an important skill that will help you understand and solve many kinds of problems, including math problems. As you read or hear a problem, you should immediately start thinking about what it means to you.

▶ Read the following problem, and think of how you would restate it in your own words.

The average family in the United States spends $100 a week on food. If the Scotts are an average American family, how much do they spend on food in 20 weeks?

You might think, "The Scotts typically spend $100 on food each week. I want to find the total they spend for food for 20 weeks."

▶ Complete the restatements for these two problems by filling in the blanks.

1. The first year a factory was in business, it employed 55 people. By the end of the second, it had grown to 105 people. How many employees did the factory add that second year?

   RESTATEMENT: A factory started out with _____ employees the first year

   and grew to _____ employees the second. I have to find out how many _____

   employees the _____ had by the _____ of that second year.

2. Trish, Marilou, Cathy, Laura, and Sharese went strawberry picking together and picked a total of 25 quarts. The friends decided to split the fruit evenly among themselves. How many quarts did each friend take home?

   RESTATEMENT: _____ friends picked 25 _____ of strawberries among

   _____, and I need to figure out _____ _____ _____ each one will get if

   they _____ the fruit _____.

▶ Now restate these problems completely in your own words. Write your ideas on a separate piece of paper. Remember to include all necessary details!

3. One Sunday, 1,500 people attended services at a neighborhood church. On the average, each person put $4 in the collection basket. How much money did the church take in that Sunday?

4. Gordon earned a gross salary of $17,300 last year. The company deducted $4,700 from his salary for federal and state taxes and for health benefits. What was Gordon's net income after deductions last year?

▶ Choose the one completely correct restatement from among the choices provided.

5. Cornelia works in the Sew Pretty fabric store. A customer recently bought an entire 125-yard bolt of material and asked Cornelia to cut it into 5-yard lengths. How many 5-yard lengths did the customer get out of the bolt?

   a. Figure how many yards of material the customer has purchased.
   b. Figure the customer's total cost for an entire bolt of material.
   c. Figure how many pieces of material the clerk can cut from the bolt of fabric.

Answers are on page 108.

Working Step by Step

If you can follow a step-by-step process, you can learn to solve problems. You can know how to do a problem, but if you aren't careful to work it out step by step, you can make mistakes. Some math problems require you to follow a certain order of operations. If you fail to follow the correct order, you can get the wrong answer.

▶ Follow the steps below to solve this problem.

$$120 \times (6 + 4) + (35 - 5 \times 2) - (15 \div 3) =$$

STEP 1. Complete the operation inside the *first* parentheses of this expression, and write the answer in the blank.

$$120 \times (\underline{\hspace{1cm}}) + (35 - 5 \times 2) - (15 \div 3) =$$

STEP 2. Do the multiplication inside the *second* parentheses. Write the answer in the blank.

$$120 \times 10 + (35 - \underline{\hspace{1cm}}) - (15 \div 3) =$$

STEP 3. Do the subtraction inside the *second* parentheses. Write the answer in the blank.

$$120 \times 10 + (\underline{\hspace{1cm}}) - (15 \div 3) =$$

STEP 4. Do the division inside the *third* parentheses. Write the answer in the blank.

$$120 \times 10 + 25 - (\underline{\hspace{1cm}}) =$$

STEP 5. Do all remaining multiplication. Write the answer in the blank.

$$\underline{\hspace{1cm}} + 20 =$$

STEP 6. Do all remaining addition. Write the answer.

This is how you should have followed the directions:

STEP 1. $120 \times (10) + (35 - 5 \times 2) - (15 \div 3) =$

STEP 2. $120 \times 10 + (35 - 10) - (15 \div 3) =$

STEP 3. $120 \times 10 + 25 - (15 \div 3) =$

STEP 4. $120 \times 10 + 25 - 5 =$

STEP 5. $1{,}200 + 20 =$

STEP 6. $1{,}200 + 20 = \mathbf{1{,}220}$

Complete the problems 1–4 with numbers that make them true. Follow these guidelines to simplify your task:

- In problem 1, start from the right and work to the left. Remember to carry a leftover amount to the next column when appropriate.

1.
```
  8 9,9 ▨ 7
+ 5 7,▨ 8 6
▨▨▨,8 6 3
```

- In problem 2, start from the right of the first multiplied line and work to the left. Then move to the next multiplied line and, finally, to the answer line, always working right to left.

2.
```
        ▨ 7 6
      × 7 ▨ 3
      ▨ 1 2 8
      ▨ ▨ ▨
   ▨ 6 3 2
   2 6 ▨,3 2 8
```

- In problem 3, start from the right and work to the left. Remember that you may have to do some borrowing from one column to another.

3.
```
   2 ▨,5 ▨ 8
 − ▨ 9,8 7 ▨
     4,▨ 2 6
```

- In problem 4, look for clues in the remainders as well as the answer line.

4.
```
            2,0 ▨ 0 r9
      2 ▨ ) 5 2,▨ 8 9
            ▨ 2
            7 ▨
            ▨ 8
              9
              0
              ▨
```

Page 28 showed an order of operations that follows this pattern:

- solve within parentheses
- from left to right—multiply or divide
- from left to right—add or subtract

Using this order of operations, solve problems 5 and 6.

5. $(14 − 2) ÷ 3 + 5$

6. $10 + (15 − 5) − 20$

Answers start on page 108.

The Five-Step Model

Some people can look at a problem and tell right away what they need to do to solve it. Many other people need a plan to get started. In the previous lessons, you practiced the skills needed to restate problems and to work in a step-by-step manner. You are now ready to put together the five basic steps that you should follow to solve any math word problem. The rest of the exercises in this chapter will help you develop your skills for all five steps.

▶ Fill in the blanks below. Use the words in the box.

> | logical   information   question   answer   operation |

1. First, you need to know what it is you are being asked to find. In other words, you must identify the _____.

2. Next, you need to figure out which numbers in the problem should be used to solve it and which, if any, are extra information. In other words, you must identify the _____.

3. Then, you must decide if you need to add, subtract, multiply, divide, or some combination. In other words, you must identify the correct _____.

4. At this point, you must do the necessary arithmetic. In other words, you must find the _____.

5. Finally, you must reread the question and make sure you have answered it logically. In other words, you must check to see that your answer is

   _____.

Let's see how we put these all together to solve a word problem.

> An ad for a clerical job says it requires a typing speed of 60 words per minute. Van takes a 5-minute test for the job and types 320 words. What is his typing speed?
>
> STEP 1. Question: **What is his typing speed?**
> STEP 2. Information: **320 words   5 minutes**
> STEP 3. Operation: + − × ÷
>   **to find words per minute you should divide (÷)**
> STEP 4. Computation:
>   **320 words ÷ 5 minutes = 64 words per minute**
> STEP 5. Check for Logic: **It is logical that if he types 320 words in 5 minutes, he will type fewer words in 1 minute.**

▶ Let's try using those steps to solve these problems.

6. The Williamses have a freezer with an inside height of 36 inches. How many frozen dinners can they stack in it if each dinner lies flat 2 inches high?

STEP 1. Question: _____   **Compute**

STEP 2. Information: _____

STEP 3. Operation: _____

STEP 4. Computation: _____

STEP 5. Check for Logic: _____

▶ Be careful with these problems. They take more than one computation to solve.

7. A game show contestant is told he will win $1 for the first 10 seconds he can keep a feather in the air and that the winnings will double every 10 seconds. How much will he win if he keeps the feather in the air for 50 seconds?

STEP 1. Question: _____   **Compute**

STEP 2. Information: _____

STEP 3. Operation: _____

STEP 4. Computation: _____

STEP 5. Check for Logic: _____

8. When Rick Mears won $804,853 in the Indianapolis 500 in 1988, he received the largest prize in auto racing history up to that point. The second-place winner got $335,103, and third place won $228,403. How much of the $5,020,000 total prize money was left for the rest of the top 33 racers?

STEP 1. Question: _____   **Compute**

STEP 2. Information: _____

STEP 3. Operation: _____

STEP 4. Computation: _____

STEP 5. Check for Logic: _____

Answers are on pages 108–109.

Predicting the Question

Have you ever looked at the cover of a book to see whether it is the type you like to read? Have you ever figured out the ending to a movie ahead of time by paying attention to the details of the plot? Have you listened to a friend's story and guessed what he or she would say next? When you did any of these things, you were making a *prediction*.

When you read a problem in math and put it into your own words, you should also be anticipating the next steps. Read the problem below.

> Kiddy World Amusement Park had a crowd-cruncher of a day with 39,575 people in attendance. The managers of the park make a profit of about $11.75 a person each day.

If you anticipated the question for this problem, you would have asked something like, **"How much profit did the park make that day?"** You can find the answer to the question by using the information given in the problem.

Predict the question for this problem:

> Juan, Roy, and Calvin opened a Speedy-Quick Oil Change franchise. By the end of the first year, they had made a profit of $33,278. They agreed to split the profit equally. How much did each partner _____?

If you finished the question with something like **"earn that first year,"** then you have the idea. The question follows from the information given in the problem.

▶ Predict the question in the next problem.

1. Judy pays $3 a pound for her favorite type of coffee at the Sav-a-Lot Food Store. She buys 5 pounds for a birthday celebration. How much does Judy

_____?

▶ Now supply the whole question for the next problem.

2. Homeville had a population of 322 in 1905. By the 1980 census the town had grown to 11,515.

_____

_____

▶ Circle the most logical question for each of the problems below.

3. Philip has a roll of string with 294 inches left on it. He needs to cut it into 7-inch-long pieces for a Cub Scout project.

    a. How long was the roll of string to begin with?
    b. How much did the roll of string cost?
    c. How many pieces will he get out of the roll?

4. The United States had a huge celebration in 1976 for its bicentennial, its 200th birthday.

    a. In what year did the United States become a nation?
    b. What country had a bicentennial the same year?
    c. What was the size of the country in 1976?

5. In a math refresher course, each student was told to buy a Whole Numbers Book for $5.75, a Decimals/Fractions Book for $6.25, and a Percents Book for $6.50.

    a. How many students were in the class?
    b. How much was each student expected to spend on books?
    c. How much were the tuition and fees for the class?

6. Look at the weather map. Write two questions based on the weather map. Make sure at least one of your questions compares two areas of the country.

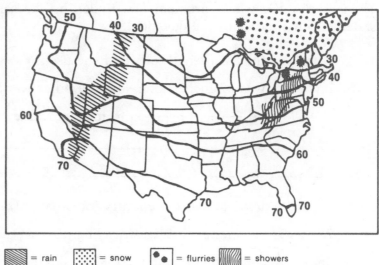

= rain   = snow   = flurries   = showers

a. _____

_____

b. _____

_____

Answers are on page 109.

Seeing the Question

To solve a problem, you have to know what the question is. In math word problems, a question, an open-ended statement, or a command sentence usually indicates the "problem." In word problems the information is given in a sequence. The three sentences in the problem below are out of order. Read each sentence and put them in order by writing 1, 2, or 3 in the box.

▪ He worked 23 hours last week.

▪ Frank makes $4.10 per hour at his part-time job.

▪ How much did he earn that week?

You should have numbered the sentences this way: (1) Frank makes $4.10 per hour at his part-time job. (2) He worked 23 hours last week. (3) **How much did he earn that week?** The question or command must logically follow the information given.

▶ Put the following word problems in sensible order by writing *1* for the sentence that should come first; *2* for the sentence that should come next, and so on. Then rewrite the whole problem and underline the sentence or open-ended statement that indicates the question.

1. ▪ What was the total length of wood she used?

   ▪ One piece was 27 inches long, and the other was 20 inches long.

   ▪ Michele used 2 pieces of wood to make a frame.

   _____

   _____

2. ▪ Jeannette drove across the country for 5 days straight.

   ▪ What was Jeannette's average traveling distance per day?

   ▪ She covered 400 miles the first day, 375 miles the next, then 390 miles, 560 miles, and finally 580 miles.

   _____

   _____

3. ■ On Monday the store received 13 boxes.

■ How many cassette tapes did the store receive that day?

■ Shipments of cassette tapes come to the Carousel Music Store in boxes of 50.

_____

_____

4. ■ What was his speed?

■ A man traveled 150 miles by car.

■ He drove for 3 hours without stopping.

_____

_____

5. ■ The tires weighed a total of 12,150 pounds.

■ How many tires were in the load?

■ He knew that all the tires weighed the same amount.

■ Mr. Johnson found that one tire from a load on his truck weighed 25 pounds.

_____

_____

6. ■ For each, the average number of years of college education is 5.

■ What is the amount spent on college by the typical teacher at Eagle Prep?

■ The teachers at Eagle Prep School are very well educated.

■ Each year of college costs about $5,600.

_____

_____

7. ■ Georgiana bought a piece of material 22 feet long.

■ How much extra material did she buy?

■ She then made dresses out of 12 feet of the material.

■ She used 6 feet to make slacks.

_____

_____

Answers are on page 109.

Labeling Information

To use numbers to solve a problem, you must know what the numbers stand for. That is why it is important to include labels with your answers. Fill in the blanks of this problem with labels that give the numbers meaning.

A weed grows at the rate of 3 _____ per week.

How many weeks will it take for the weed to grow 15 _____?

You probably filled in the blanks with the word **inches** (or feet if you have monstrous weeds in your part of the country!). The problem makes a lot more sense with the labels put in.

▶ Fill in the blanks in these problems with sensible labels.

1. Sandra had 400 _____. She spent 80 _____. How much did she have left?

2. A nurse worked 36 _____ one week, 20 _____ the next, and 32 _____ the week after that. How long was his average weekly work schedule?

3. Sam ate 2 bananas. If each banana weighed 9 _____, how many _____ of bananas did Sam eat?

4. There are four trucks at a truck stop. One truck driver bought 20 _____ of gas, another 32 _____, the third 19 _____, and the fourth 41 _____. How much gas did they buy?

When you solve a problem, write down the numbers with their labels. Be sure you know what label should go with your answer.

A clerk at Desmond's Department Store earns $5 per hour. If one of the clerks worked 39 hours one week, how much did she earn?

INFORMATION: **$5** per hour	ANSWER LABEL: $
39 **hours**	
SOLUTION: $5 × 39 hours = **$195**	

Even if you had multiplied correctly, 195 *hours* would be the wrong answer. Be sure to always reread the question and write the correct label on the answer.

▶ In the following problems, pick out the numbers you need in order to find the solution. Write the numbers and their labels under the heading "Information." Then solve the problem and circle the label in the answer.

5. Of 183 students who enrolled in a night school program one year, only 98 successfully completed their classes. How many were *not* successful?

INFORMATION: SOLUTION:

_____   _____

6. A book has 410 pages. Each page cost 38 cents to print. How much did it cost to print the book?

INFORMATION: SOLUTION:

_____   _____

7. If a train covered 720 miles in 12 hours, how fast was it going?

INFORMATION: SOLUTION:

_____   _____

8. Iris sold 31 pounds of tomatoes at her roadside stand one weekend. Jane sold 57 pounds of tomatoes that same weekend. How much more did Jane sell than Iris?

INFORMATION: SOLUTION:

_____   _____

Answers are on page 109.

Sorting Out Information in Word Problems

Why do word problems have **unnecessary information**? Actually problems with extra information resemble real life. Often we have more information than we need to solve a problem, and we have to choose only the necessary information.

One thing that makes word problems hard to solve is that they may contain more information than you need. Look at the example below:

Patsy bought two 32-ounce jars of mayonnaise on sale last week. She saved 40 cents on each jar. How much did she save on the mayonnaise altogether?

STEP 1. What is the question?
How much did she save?

STEP 2. What do I need in order to find that? (*Write the numbers and the labels.*)
two jars
40 cents saved on each

STEP 3. What is *not* needed to solve the problem?
32-ounce jars

Why?
The size of the jars is not needed to solve the problem.

▶ Included in the following problems is unnecessary information. Circle the letter identifying information that you *don't* need in order to answer the question. *Do not solve the problems yet.*

1. A chair is on sale at Frank's Furniture Mart. It originally sold for $560 and is now being advertised for $390. Frank's employees can get an additional $80 discount. How much would one of Frank's employees pay for the chair on sale?

It is *not* necessary to know:

a. $560 original price
b. $390 sale price
c. $80 discount

38

2. The Sampsons have the Cleveland *Plain Dealer* delivered to their home every day but Sunday for 30 cents per day. Every Sunday they buy the New York *Times* for $1.00. During one 30-day, 4-weekend month, how much did they spend on the *Plain Dealer*?

It is *not* necessary to know:

a. 30-cents-per-day cost for the *Plain Dealer*
b. $1.00-per-Sunday cost for the *Times*
c. 30-day, 4-weekend month

3. An earthquake took the lives of 464 people in one village. 54 people were hospitalized, and 28 were treated and released. 109 people were uninjured but lost their homes. How many people were injured or killed in the disaster?

It is *not* necessary to know:

a. 109 lost their homes
b. 28 were treated
c. 464 were killed
d. 54 were hospitalized

4. The local night school offered 21 GED classes in 5 different locations last semester. If each class had an average enrollment of 12 women and 9 men, how many GED students did the school enroll last semester?

It is *not* necessary to know:

a. 21 classes
b. 12 women
c. 9 men
d. 5 locations

▶ Now solve the problems above, using only the necessary information.

5. $390.00
 $- \ \ 80.00$

7. 464
   54
 $+ \ 28$

6.   26
 $\times .30$

8.   12   21
 $+ \ 9$ $\times 21$

Answers are on page 110.

Recognizing Incomplete Information

Believe it or not, some word problems do not have enough information to solve them. Is this just a mistake? No. In some cases, you are being tested to see whether you understand what is needed to solve a problem. Look at the following example.

> Handy Man's is selling stockade-style fencing in 6-foot sections. How many sections do the Jamisons need to purchase to enclose their yard?

Is enough information given to solve the problem? **no**
If not, what is missing? **the size of the Jamisons' yard**

▶ In the following problems, identify whether enough information is given to enable you to reach a solution. If more information is needed, tell what that is.

1. Jane's boss told her to file 200 documents. She can file 50 documents per hour. How many hours will it take her to finish the task?

   Is enough information given to solve the problem? _____

   If not, what is missing? _____

   _____

2. Thomas's old clunker uses 15 gallons of gas a week. Thomas drives to and from work 5 days a week. What mileage (miles per gallon) is Thomas getting with his car?

   Is enough information given to solve the problem? _____

   If not, what is missing? _____

   _____

3. Christy bought a new washing machine. She put $50 down and agreed to pay $30 a month on an installment plan. How much did she pay for the washing machine?

   Is enough information given to solve the problem? _____

   If not, what is missing? _____

   _____

4. Bart found 68 empty pop cans in his apartment the morning after one of his parties. There were 16 unopened cans still in his refrigerator. How much pop had he figured he needed per guest?

Is enough information given to solve the problem? _____

If not, what is missing? _____

_____

5. Main Street Office Supplies ordered 17 cases of manila envelopes from its supplier. Each case contains 12 boxes of envelopes. How many boxes of envelopes were ordered?

Is enough information given to solve the problem? _____

If not, what is missing? _____

_____

6. Look at the windows below. The windows are 6 feet high and 4 feet wide. How much will curtains cost to cover the windows?

Is enough information given to solve the problem? _____

If not, what is missing? _____

_____

Answers are on page 110.

Recognizing Incomplete Information in Problems

On tests you may be given a choice of answers, and you will have to choose the correct one. Occasionally, you will not be able to solve the problem because not enough information is given. In these cases, you may have to:

1. Choose the answer *not enough information is given.*

OR

2. Tell what information is needed to solve the problem.

Look at the example below:

> There are 279 people working in a 9-year-old office building. On the average, how many people work on each floor?
>
> **a.** 31 people per floor
> **b.** 2,511 people in the building
> **c.** We need to know how many floors.
> **d.** We need to know how many offices per floor.

STEP 1. Question: **How many people work on each floor? (average)**
STEP 2. Information: **279 people**
STEP 3. Compute: **279 people ÷ number of floors = average**

The correct answer is **c. need to know the number of floors**. Since there is not enough information here, you can't solve the problem.

WARNING: Don't just divide 279 people by the number 9 from the 9-year-old building. Choice *a—31 people*—may seem right because 279 ÷ 9 = 31, but don't fall into the trap. The age of the building is not important in this problem.

▶ Now complete the problems by either finding the correct solution—from *a* or *b*—or indicating what information is needed—from *c* or *d*.

1. What would be the cost for a family to buy the following school supplies: 2 pairs of scissors at $3 each, 4 tablets of paper at $2 each, and 6 pens at $1 each?

    a. $6 for the supplies
    b. $20 for the supplies
    c. We need to know how many school children are in the family.
    d. We need to know what supplies the school provides.

2. A truck driver covers 495 miles one day, 605 miles the next day, and the rest of the trip to Dubuque on the third day. How far did the trucker have to travel to get to Dubuque?

    a. 1,100 miles
    b. 110 miles
    c. We need to know how far he went on the third day.
    d. We need to know where the trucker started his trip.

3. Mr. Scott and Mr. Spingola set up a community hall for a speaker's night. They put 15 rows of 30 chairs on one side of an aisle and the same arrangement on the other side of the aisle. How many people could this hall then seat?

    a. 450 people altogether
    b. 900 people altogether
    c. We need to know how many speakers there would be.
    d. We need to know how many people had tickets for the event.

4. Michael cut two pieces of lumber 4 feet long each, three pieces 3 feet long each, and two pieces each 1 foot in length. How much more did he need to cut to make a desk?

    a. 19 feet altogether
    b. 8 feet more
    c. We need to know how much lumber was wasted in the first cuts.
    d. We need to know how much lumber he needs to build the desk.

5. Look at Delma's idea of a healthy lunch. This is what she ate every day while she did temporary work in a downtown office. The hamburger had 220 calories, the fries had 180 calories, and the cola had 120 calories. How many calories did she consume at lunchtime while working at that office?

    a. 520 calories
    b. 1,560 calories
    c. We need to know how much she threw away.
    d. We need to know how many days she ate that lunch.

Answers are on page 110.

Identifying the Operation

Being able to identify what operation is needed to solve math problems is essential to the problem-solving process.

▶ In the situations described below, fill in the operation that would be necessary to find a solution.

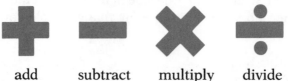

add   subtract   multiply   divide

1. You _____ to figure how many calories you ate altogether one day.

2. You _____ the total of the bills you have to pay from your weekly paycheck amount to figure if you can afford dinner and a show this week.

3. You _____ to find the total cost of several cans of the same paint for a redecorating project.

4. You _____ to figure the cost for one can of pop when you have bought a whole case of it.

5. You _____ to find out how much you will earn cleaning houses when you charge one flat rate for a cleaning.

6. You _____ to find how many boxes of Girl Scout cookies you bought if you can remember the total money you owe and the cost for each box.

7. You _____ the total number of men from the total number of women in the United States to find out how many more women than men there are.

8. You _____ the amount of each check you write from the previous balance to find the new balance.

Answers are on page 110.

Rounding Numbers to Choose Operations

Some word problems seem difficult because the numbers in them are so large that they seem threatening. In such problems, you may be able to see the patterns within them, if you don't get confused by the big numbers. Sometimes you can **round** the numbers to make the problem easier to understand. Then you can decide which operation to use. When you round, you will have to decide whether you want to round to the nearest ten, hundred, or thousand. Remember, if you round one number in a problem, you must round them all.

Look at the example below.

> A pilot flew 1,537 miles during one stint of 3 hours and 15 minutes.
> At what speed (mph) was he flying?

Now reread the problem with the simpler numbers 1,500 and 3. You can see that this is clearly a division problem.

$$1,500 \text{ miles} \div 3 \text{ hours} = 500 \text{ mph}$$
$$\text{so}$$
$$1,537 \text{ miles} \div 3 \text{ hours } 15 \text{ minutes} = 473 \text{ mph}$$

▶ In the following problems, simplify the numbers by rounding them to make the patterns of the problems easier to see. Then solve using the rounded numbers. Check your work by solving the original problems on another sheet of paper.

1. A dictionary sells for $65. Approximately how many of these dictionaries can an adult education program buy with a grant for $9,570?

2. Steve Farmer owns 1,526 cows. This is 498 fewer than Joe Rancher owns. Approximately how many cows does Joe own?

3. A certain town has 2,786 households. If each household has an average of 22 small appliances, approximately how many small appliances are there altogether in the households of this town?

4. A school purchased 1,652 yards of lumber for projects in its woodworking classes. If the students used 1,397 yards of it, approximately how much lumber was left at the end of the year?

5. A dense object falls from a plane at the rate of 76 yards per second. In 53 seconds, approximately how far will it fall?

Answers are on page 110.

Comparing Problems to Choose Operations

It is helpful to be able to see similarities between word problems. If you can recognize that a problem is similar to one you have worked before, then you will find it easier to set up and solve.

See if you can recognize which two problems out of three have the same pattern. Be careful! Similar information given in two problems does not necessarily mean the same operation will be used to solve them. Look at the example below.

a. Elwood drives back to his home town every weekend. If the round trip is 350 miles, how many miles will he have traveled in 4 weekends?

OPERATION: _____

b. Jonathan and his girlfriend go to different schools. Their schools are 980 miles apart. If Jonathan has traveled 1,960 miles round trip in the last year to visit his girlfriend, how many times has he visited her?

OPERATION: _____

c. Tory goes to the store and back every day. If he goes 5 miles each day, how far does he go in 30 days?

OPERATION: _____

Choices *a* and *c* are similar because they follow the pattern of **multiplying to find a total amount**. Choice *b* is a division problem.

▶ Read each problem, and write the operation you would use to solve it. Then complete the sentence that explains the similarity between two of the problems.

1. a. A consumer's magazine rated 12 cars in one issue. All but 2 were foreign cars. How many domestic cars did the magazine rate?

OPERATION: _____

b. There are 23 sports teams at one college. Three of the teams are for females, the rest are all male. How many teams at the college are for men?

OPERATION: _____

c. At a small bank, there are 15 clerks who have high school diplomas, 7 who have completed some college, and 6 who have college degrees. How many clerks at the bank have at least a high school education?

OPERATION: _____

Choices _____ and _____ are similar because they follow the pattern of

_____.

2. a. Franklin makes $12 per hour. How much would his gross pay be after 23 hours of work?

OPERATION: _____

b. Eggs are sold at a farmer's stand for 5 cents each. How much do a dozen eggs cost at the stand?

OPERATION: _____

c. A package of gum costs 30 cents. If there are 6 sticks in a package, how much does each stick cost?

OPERATION: _____

Choices _____ and _____ are similar because they follow the pattern of

_____.

3. a. If a box of chocolates weighs 16 ounces, and each piece of candy weighs 2 ounces, how many pieces are in the box?

OPERATION: _____

b. If three friends equally share a box of hard candies, and each one eats 5 ounces, how many ounces of candy were in the box?

OPERATION: _____

c. If a box of gumdrops contains 330 calories, and each gumdrop contains 6 calories, how many gumdrops are in the box?

OPERATION: _____

Choices _____ and _____ are similar because they follow the pattern of

_____.

Answers are on page 110.

Estimating Answers to Word Problems

Sometimes an ***estimated answer*** is all you need—or all you have time to get—to solve a problem. Estimating is also a quick way to check for bigger, more obvious errors in figuring. On page 14, you learned to round numbers. Look at the example below, which uses rounded numbers to estimate answers, and compare these answers to the real solution.

> Pacelak Grocery Store receives mail 313 days a year. On the average, the store gets 17 pieces of mail each day. How much mail does it receive in a year?
>
> ESTIMATED INFORMATION: 300 days, 20 pieces of mail per day

Estimated Answer	Real Answer
300 days × 20 pieces 6,000 pieces	313 days × 17 pieces 5,321 pieces

▶ Now do each problem below. First find an estimated answer, then compare it to the real answer.

1. Belinda buys all her greeting cards from a mail-order firm that charges 13 cents less per card than card stores do. Since Belinda buys 147 cards each year, how much does she save during that time through buying by mail?

   ESTIMATED INFORMATION: _____

Estimated Answer	Real Answer

2. Sally read a 468-page book in a total of 18 hours. How many pages did she read per hour?

ESTIMATED INFORMATION: _____

Estimated Answer	Real Answer

3. A sofa that usually sells for $871 is on sale for $199 less this week. How much is it on sale for this week?

ESTIMATED INFORMATION: _____

Estimated Answer	Real Answer

4. During the last year, the Nortons have charged the following amounts on various credit cards: $1,320 for a vacation, $229 for a new dryer, $1,898 for a fence around their property, and $2,102 for miscellaneous clothes and household goods. How much did the Nortons charge during the last year?

ESTIMATED INFORMATION: _____

Estimated Answer	Real Answer

Answers are on page 111.

Deciding Whether the Answer Makes Sense

After finding the answer to a problem, you should take the time to check to make sure your solution answers the question and makes sense. Often you can catch a careless mistake and correct your work. Look at the example below. Read the problem and write *L* next to the answers that seem logical.

Minnie averages $57 a night in tips at her waitressing job. How much has she earned in tips after 36 nights at her job?

_____ **a.** $100

_____ **b.** $240

_____ **c.** $2,100 (*by rounding $57 to $60 and 36 nights to 35 nights*)

_____ **d.** $2,400 (*by rounding $57 to $60 and 36 nights to 40 nights*)

You should have written *L* next to choices c and d. The other two choices are *not* logical since Minnie can earn $114 in two nights and nearly $240 in four nights.

HINT: Estimating answers can help you to see if your answer makes sense, but only by calculating carefully can you be sure that your answer is correct.

▶ In the following problems, identify each of the answer choices that seems close to the actual answer by putting an *L* (for logical) on the line next to it. In some cases below, a problem may have more than one logical answer. Use your skill with estimating to help you.

1. If the population of the United States was 3,929,214 in 1790 and gained 1,379,269 people within 10 years, what was the nation's population in 1800?

   a. _____ 3,500,000   **c.** _____ 5,000,000

   b. _____ 4,000,000   **d.** _____ 5,500,000

2. Duwayne is a long distance truck driver. He covered 68,109 miles last year. If he drove 219 days that year, how many miles did he drive on the average per day?

   a. _____ 310 miles   **c.** _____ 3,000 miles

   b. _____ 340 miles   **d.** _____ 3,500 miles

3. The Morellis think that they can sell their present home for $81,950. They have their eyes on a new home that would cost them $120,575. How much is the price difference between the homes?

a. _____ $4,000

c. _____ $140,000

b. _____ $40,000

d. _____ $200,000

4. Twenty-eight churches in one county agree that each will house up to 118 homeless on each night that the temperature dips below 20 degrees during the winter months. What is the maximum number of homeless that will be taken in on any cold winter night in these churches?

a. _____ 36,000

c. _____ 3,000

b. _____ 3,600

d. _____ 360

5. Compare the heights of the two buildings at the right. Approximately how much taller is the Sears Tower than the Standard Oil Building?

    a. 40 feet
    b. 400 feet
    c. 2,500 feet
    d. 2,600 feet

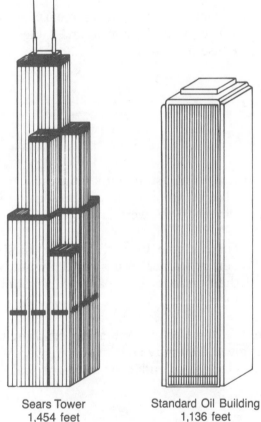

Sears Tower
1,454 feet

Standard Oil Building
1,136 feet

Answers are on page 111.

4 Measurements

If you have ever bought carpeting or tiles for a floor, built a bookcase, cooked a dinner, served as a timekeeper, or figured a distance, you have worked with measurements. Understanding measurements can make your life easier at home, on the job, and in the classroom.

In this chapter you will learn to:

- compare and convert measurements
- use pictures, charts, and tables
- use formulas

Comparing Measurements

To work with measurements, you must be able to get a picture in your mind of the size or amount they represent. One way to do this is to **compare** a measurement to an everyday object.

> For example, an inch is closer in size to a paper clip than to a pen or stapler. You already know that a pen or stapler is longer than an inch, so a paper clip would be the object closest to an inch in size.

▶ From the items below, underline the thing that is closest in size to the measurement given. You may check your answers with a ruler, a yardstick, a measuring cup, a clock, a calendar, or a weight scale.

1. a foot:

   a. the length of a phone book
   b. the edge of a desk
   c. the edge of a postcard

2. a yard:

   a. a Ping-Pong paddle
   b. a tennis racket
   c. a baseball bat

3. a fluid ounce:

    a. the amount of liquid in a raindrop
    b. the amount of liquid in a bottle of nail polish
    c. the amount of liquid in a can of soup

4. a cup:

    a. the amount of detergent to wash a load of dishes
    b. a serving of coffee in a restaurant
    c. the amount of water in an aquarium

5. a gallon:

    a. a can of soda pop from a machine
    b. a bottle of ketchup at a restaurant
    c. the largest container of milk from the store

6. a second:

    a. the time it takes to say your first name
    b. the time it takes to sing a song
    c. the time it takes to brush your teeth

7. a week:

    a. the time from one electric bill to the next
    b. the time from one episode of a TV series to the next
    c. the time from one brushing of your hair to the next

8. a pound:

    a. a jar of spices
    b. a piece of cake
    c. a loaf of bread

▶ Circle which item in each of the following pairs is bigger. The first is done for you as an example.

  9. foot   (yard)

10. decade   century

11. ton   pound

12. cup   pint

13. gallon   quart

14. meter   kilometer

15. teaspoon   tablespoon

Answers are on page 111.

Converting Measurements

Now you have an idea of the size of some basic measurements. You will find many occasions to apply measurements to solving problems. To do so, you have to be able to **convert** from one size to another by multiplying or dividing. The following chart will help you make the conversions.

Standard U.S. Measurements		
Time	**Length**	**Weight & Capacity**
60 sec = 1 min	12 in = 1 ft	16 oz = 1 lb
60 min = 1 hr	36 in = 1 yd	2,000 lb = 1 T
24 hr = 1 day	3 ft = 1 yd	8 fl oz = 1 c
7 days = 1 wk	5,280 ft = 1 mi	2 c = 1 pt
12 mo = 1 yr	1,760 yd = 1 mi	2 pt = 1 qt
365 days = 1 yr		4 qt = 1 gal
366 days = 1 leap yr		
10 yr = 1 decade		
100 yr = 1 century		

Looking at the chart, you see that a yard is a larger unit than a foot. To convert a larger unit of measure to a smaller unit, you **multiply**. This is because you want **more** of the smaller unit. To convert a smaller unit of measure to a larger unit, you **divide**. This is because you want **fewer** of the larger unit. Follow the steps below to decide whether to multiply or divide.

6 feet = ? yards

Is a foot a larger or smaller unit than a yard? **smaller**

Are you converting a smaller unit to a larger unit? **yes**

Do you multiply or divide? **divide**

So, 6 feet ÷ 3 feet = **2 yards.**

HINT: First decide whether you are converting a larger unit to a smaller unit or a smaller unit to a larger one. It helps to write *sm* above the smaller unit and *lg* above the larger one.

▶ Now complete the following problems, showing your work in the space provided.

1. **a.** 3 tons = _____ pounds   Compute
   b. multiply or divide?

2. a. 3 minutes = _____ seconds   **Compute**
   b. multiply or divide?

3. a. 32 ounces = _____ pounds
   b. multiply or divide?

4. a. 60 inches = _____ feet
   b. multiply or divide?

5. a. 364 days = _____ weeks
   b. multiply or divide?

6. a. 8 decades = _____ years
   b. multiply or divide?

7. a. 48 hours = _____ days
   b. multiply or divide?

8. a. 3 cups = _____ fluid ounces
   b. multiply or divide?

9. a. 3 miles = _____ feet
   b. multiply or divide?

10. a. 24 months = _____ years
   b. multiply or divide?

Answers are on page 111.

Drawing Pictures

Sometimes a measurement problem is easier to solve when you can **picture** it. People who are good at solving problems often take the time to sketch out the problem and put the information in the picture.

Read the following problem, and look at the picture that has been drawn showing its information.

The Vegas are fencing in their backyard. The fence will extend 8 feet from each side of the house and then go 98 feet to the end of the yard and 210 feet across the back. How much fencing do the Vegas need?

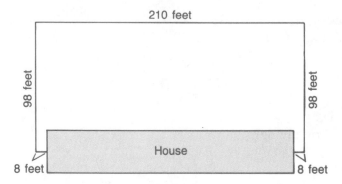

It is clear after drawing and labeling the information from the problem that you must add together the two 8-foot sides, the two 98-foot sides, and the one length of 210 feet.

$$2(8 \text{ ft}) + 2(98 \text{ ft}) + 210 \text{ ft} =$$
$$16 \text{ ft} + 196 \text{ ft} + 210 \text{ ft} = \textbf{422 ft}$$

▶ First draw pictures of the following problems for help in figuring them out. Be careful. Some are one-step and some are multistep problems. Then solve the problems.

1. Lester has agreed to make a frame for the wall-hanging his wife has cross-stitched. The piece is 11 inches by 12 inches. How much wood does he need to buy?

    Draw Here

2. Gail bought a piece of material 48 inches wide and 3 yards long. Out of it she cut 2 curtains, each requiring 48 inches in width and 4 feet in length. How much material did Gail have left?

Draw Here

3. Ernie is building a home with 8 doors that need framing. Each door is 3 feet wide and 9 feet high. How much framing does he need? (*When you draw your picture, think carefully about where the framing goes!*)

▶ In the drawing below, write the measurements on the lines and solve the problem.

4. The Fontanellas decide to put sod in their front yard. Their property is 92 feet across and 53 feet up to the border that fronts the house. Their driveway, which crosses the yard, is 12 feet wide and 53 feet long. How much sod must they buy?

Label Here

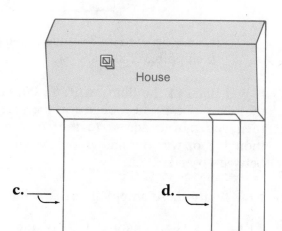

House

c. ____

d. ____

a. ____

b. ____

Answers are on page 111.

Getting Information from Charts and Tables

Information needed for some measurements is often given in a **chart** or on a **table**. For example, the following chart would be used to figure shipping costs on a catalog order.

Weight	Shipping Cost
5 oz or less	$ .50
6–10 oz	$1.00
11–15 oz	$1.50
1–2 lb	$2.50
2 lb 1 oz–5 lb	$5.00
over 5 lb	$7.50

If Susie ordered an 8-ounce bottle of perfume, she paid _____ in shipping charges.

If you filled in the blank with **$1.00**, you know how to get information from the chart. First you looked for the given amount in the weight column, and then you moved across to the cost column to find the correct figure. Even though 8 oz was not shown on the chart, you can see that it falls in the range between 6 and 10 oz.

▶ Now figure these shipping costs, using the chart above.

1. Larry ordered a pair of shoes that weigh 1 lb 7 oz. He paid _____ in shipping charges.

2. Colin got 2 books that weighed 20 oz each. He paid _____ for shipping costs.

3. Daniella bought a skirt weighing 1 lb, a blouse weighing 13 oz, and pantyhose weighing 4 oz. She paid _____ for shipping charges.

4. Harvey purchased a 4-oz bottle of cologne, a 2-oz container of scented powder, and a 1-oz tube of hair gel. He spent _____ on shipping costs.

5. Erline ordered a set of sheets that weighed 2 lb 10 oz, and a set of towels that weighed 3 lb 6 oz. She totaled _____ in shipping costs.

The same catalog supplies several tables to help you choose your correct size. Answer the next set of questions based on two of those tables.

Women's Regular—fits 5′4″ to 5′7″

	X-Sm	Small		Medium		Large		X-Lg
	4	6	8	10	12	14	16	18
Bust	32	33	34	35	36	38	40	42
Waist	24	25	26	27	28	30	32	34
Hips	35	36	37	38	39	41	43	45
Sleeve	28	29	29	30	30	31	31	32

Women's Petite—fits 4′11″ to 5′3″

	X-Small	Small		Medium		Large
	4	6	8	10	12	14
Bust	32	33	34	35	36	38
Waist	24	25	26	27	28	30
Hips	35	36	37	38	39	41
Sleeve	26	27	28	28	29	29

6. Tory wants to order a top for his girlfriend. She is 5 feet 2 inches tall and her bust line measures 35 inches around. Should he order a petite or regular size? _____

   Should he order x-small, small, medium, or large? _____

7. Grace is 5 feet 6 inches tall and has a 34-inch waist and 45-inch hips. Does she need to order pants in regular or petite? _____ Which size should she order? _____

8. If Lynette is 5 feet 5 inches tall and orders size 14 in dresses, her measurements are probably about _____, _____, _____, and _____.

9. Mona is close to 5 feet 3 inches tall, and she has a 34-inch bust and a 28-inch sleeve length. What size should she order? (*Be sure to specify regular or petite.*) _____.

Answers are on page 112.

Multistep Measurement Problems

Just as with other problems in life, measurement and other math problems often involve more than one step. It helps to break a multistep problem down into smaller problems. Look at the example below.

The Ozogs have a lot that measures 200 feet by 200 feet. They need to cover the area with seed that costs $8 per box, covering 2,000 square feet. How much will the grass seed cost them?

QUESTION: How much will the seed cost?

Problem 1	Problem 2	Problem 3
QUESTION: What is the area to be covered?	QUESTION: How many boxes are needed to cover 40,000 square feet?	QUESTION: How much will 20 boxes of seed cost?
$200 \times 200 =$ 40,000 square feet	$40,000 \div 2,000 =$ 20 boxes	20 boxes $\times$ $8 = $160

▶ The next two multistep problems are based on the information in the chart below. For both of the problems, solve each mini-problem.

Food Item	Calories
8 ounces coffee	0
16 ounces diet pop	2
1 cup whole milk	150
1 cup low-fat milk	120
1 ounce butter	50
4 ounces fish	80
8 ounces cottage cheese	215
1 pint yogurt	240
1 pint strawberries	100

1. Lou had a cup of coffee for breakfast. For lunch, he had a cup of whole milk, 4 ounces of fish cooked in 1 ounce of butter, 8 ounces of cottage cheese, and a half-pint of strawberries. He is allowed 2,500 calories a day on his diet. How many calories can he have for dinner?

Problem 1	Problem 2	Problem 3
QUESTION: How many calories are in a half-pint of strawberries?	QUESTION: How many calories did Lou consume so far?	QUESTION: How many calories can Lou eat at dinner to reach 2,500?

2. Clare had 16 ounces of diet pop, a cup of cottage cheese, and a half-pint of strawberries for lunch. Jodi had 4 ounces of fish cooked in an ounce of butter and 1 cup of low-fat milk. Which woman had more calories for lunch?

Problem 1	Problem 2	Problem 3	Problem 4
QUESTION: How many calories are in a half-pint of straw-berries?	QUESTION: How many calories did Clare consume?	QUESTION: How many calories did Jodi have?	QUESTION: Which woman had more calories for lunch?

▶ First read the entire problem and decide what two mini-problems need to be solved. Then write the question for each mini-problem. Finally, solve the whole problem.

3. Robyn had 2 cups of coffee and a half-pint of yogurt for breakfast. For lunch she had 8 ounces of cottage cheese, a half-pint of strawberries, and an 8-ounce glass of diet pop. For dinner she had 6 ounces of fish broiled without butter, 8 ounces of cottage cheese, and 16 ounces of diet pop. Her diet allows 1,200 calories a day. How far from her allowance was she?

Problem 1	Problem 2
QUESTION:	QUESTION:

Answers are on page 112.

Using Formulas

Many **formulas** have been developed to make work with measurements easier and faster. Many common objects are rectangular, so it is helpful to know three special formulas.

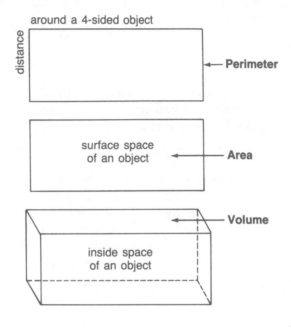

Perimeter of a rectangle =
Two lengths + Two widths
(P = 2*l* + 2*w*)
Answers given in units.

Area of a rectangle =
Length × Width
(A = *lw*)
Answers given in square units.

Volume of a rectangle =
Length × Width × Height
(V = *lwh*)
Answers given in cubic units.

The Dixons want to fence in their backyard, which is 85 feet wide and 135 feet long. How much fencing do they need?

STEP 1. Question: How much fencing do they need?
STEP 2. Formula: Since I need the distance around, I choose perimeter. *P = 2l + 2w*
STEP 3. Necessary Information: 85 feet wide, 135 feet long
STEP 4. Solution: *P* = 2(135) + 2(85) = **440 ft**
STEP 5. Check for Logic: 440 feet seems reasonable for fencing a piece of property this size.

▶ In the following problems, use the 5-step model to find the solution. To determine the operation, you will have to choose the correct formula.

1. George needed litter for his cat's box. The cat's litter box is 20 inches wide, 32 inches long, and 4 inches high. How many cubic inches of litter does George need to buy to fill the box halfway?

   STEP 1. Question:

   STEP 2. Formula:

   STEP 3. Necessary Information:

STEP 4. Solution:

STEP 5. Check for Logic:

2. Lilli Mae is sewing a rectanguar tablecloth around which she will attach lace. If the cloth is 2 yards wide and 3 yards long, how much lace does Lilli Mae need to buy?

STEP 1. Question:

STEP 2. Formula:

STEP 3. Necessary Information:

STEP 4. Solution:

STEP 5. Check for Logic:

3. Victor has made a picture frame and needs glass to cover the picture inside. The inside dimensions of the frame are 8 inches wide by 13 inches long. How large should Victor's piece of glass be?

STEP 1. Question:

STEP 2. Formula:

STEP 3. Necessary Information:

STEP 4. Solution:

STEP 5. Check for Logic:

4. The Pro Sport Shop receives golf balls packaged in 4-pack containers that measure 4 inches long, 2 inches wide, and 2 inches high. The 4-pack containers are shipped inside boxes 11 inches high, 12 inches wide, and 16 inches long. How many containers of balls come shipped in a full box?

STEP 1. Question:

STEP 2. Formula:

STEP 3. Necessary Information:

STEP 4. Solution:

STEP 5. Check for Logic:

Answers are on page 112.

2 Decimals, Fractions, and Percents

Knowing how to work with money, how to measure, and how to interpret data about people is necessary in today's world. Decimals, fractions, and percents are all numbers that concern the relationships of parts to wholes. We are exposed to decimals, fractions, and percents whenever we make cash purchases, distribute items evenly, or receive facts about groups of people within the larger society. Much of the time we have to use these numbers in problem situations. In this section, you will learn strategies to solve problems involving decimals, fractions, and percents.

5 Decimals

Every time you use money in the United States, you use decimals. That makes the ability to understand decimals very important. But it also makes understanding decimals easier because you use them all the time.

In the world of measurements, decimals are vital. Decimals are used in standard U.S. measurements to express parts of wholes. Metric measurements are based entirely on decimal units. Every year in this country the metric system becomes more widely used and accepted.

In this chapter, you will learn to:

- estimate and compare with decimals
- use metric measurements
- use decimals in working and shopping situations

Knowing When to Estimate

Sometimes you are in a situation in which a problem does not need an exact answer. Have you ever looked in your checkbook to see if you had enough to make a purchase and found (oh, no!) that you had not yet subtracted your last three checks from the balance? You probably did a quick mental check of the figures by *estimating* as in the situation below.

CHECK #	DATE	ISSUED TO	AMOUNT	BALANCE $267.98	close to $270
1023	7/2	Jolly Foods	$63.29		close to $60
1024	7/2	Our Pet Supplies	$17.67		close to $20
1025	7/3	Horizon Hair Cuts	$21.25		close to $20

You can estimate:

$$\$60 + 20 + 20 = \$100$$
$$\$270 - 100 = \$170$$

You know that you can spend about **$170**.

Although estimates are useful, you must find an exact answer when you balance your checkbook. On another sheet of paper, subtract the amounts to find the *exact* balance. Was your estimated answer close?

▶ Identify which of the following situations absolutely require an exact answer and which could be handled with an estimate. Use a check mark.

1. A manager needs to compare the profits his store earned during a month of heavy advertising with the profits earned during a month of in-store promotions.

    _____ exact   _____ estimate

2. A high school student is saving money for college and wants to figure how much more he needs for his first two years.

    _____ exact   _____ estimate

3. A machinist is making a part that must fit snugly inside another part.

    _____ exact   _____ estimate

4. An expectant mother is budgeting money for disposable diapers for the first month.

    _____ exact   _____ estimate

5. A weather reporter is figuring out whether or not the amount of rain that fell in April broke a record.

    _____ exact   _____ estimate

6. A tailor is measuring buttonholes for a suit he is making.

    _____ exact   _____ estimate

7. A plant lover is buying pounds of potting soil to repot her dozen or so plants at home.

    _____ exact   _____ estimate

8. A timekeeper is deciding which runner had the shortest time in the Olympic trials.

    _____ exact   _____ estimate

Answers are on page 113.

Estimating Answers

At some time, you probably have gone shopping with a limited amount of money and have had to be careful not to spend too much. Skill with rounding and estimating numbers can help you in such a situation.

Martha went to the store with just a $20 bill in her purse. She picked out earrings for $7.99, perfume for $9.25, and nylons for $1.75. Excluding taxes, will Martha have enough money?

In estimating,
$7.99 becomes $8.00 ($.99 is greater than $.50, or half a dollar)
$9.25 becomes $9.00 ($.25 is less than $.50, or half a dollar)
$1.75 becomes $2.00 ($.75 is greater than $.50, or half a dollar)
　　　　　　$19.00 is the estimate

So, yes, Martha has enough. As a check of your estimate, add the original numbers. You should get the answer $18.99, which is very close to the estimate.

▶ Round the numbers in the problems and estimate the answers. Use the five-step method. Watch for multistep problems. Then do the actual calculations to check your logic. The first one is done for you.

1. Mort has a balance of $171.19 in his checkbook. He writes a check for $39.88 to the electric company and $26.14 to the phone company. What is his new balance? (*Round to the nearest ten dollars.*)

   a. QUESTION: What is Mort's new balance?

   b. NECESSARY INFORMATION (*estimated*): $170, $40, and $30

   c. NUMBER SENTENCE: $170 − ($40 + $30) = ?

   d. ESTIMATE: $170 − $70 = **$100 estimate**

   e. CHECK (*go back and do the steps with exact figures*):
   $171.19 − ($39.88 + $26.14) = **$105.17 exact**
   HINT: Notice that you can write the mini-problem in parentheses as in the number sentence above.

2. Sally makes cuts of .290 inches, .413 inches, and .651 inches off the end of a piece of metal that was 2.5 inches long. How long was the piece after Sally made the three cuts? (Round to the nearest tenth to estimate.)

   a. QUESTION:

   b. NECESSARY INFORMATION (*estimated*):

   c. NUMBER SENTENCE:

   d. ESTIMATE:

e. CHECK (*go back and do the steps with exact figures*):

3. A butcher sold 27.28 pounds of pork chops at $2.79 a pound. How much money did he take in for pork chops that day? (Round to the nearest whole number to estimate.)

   a. QUESTION:

   b. NECESSARY INFORMATION (*estimated*):

   c. NUMBER SENTENCE:

   d. ESTIMATE:

   e. CHECK (*go back and do the steps with exact figures*):

4. When Ellie works in the gift-wrapping department, she precuts ribbon into 37.5-inch-long pieces. How many pieces can she cut from a roll 759.5 inches long? (Round to the nearest whole number to estimate.)

   a. QUESTION:

   b. NECESSARY INFORMATION (*estimated*):

   c. NUMBER SENTENCE:

   d. ESTIMATE:

   e. CHECK (*go back and do the steps with exact figures*):

5. Benjy "the Bunter" Boynton had a .297 batting average his first year. The second year, it was .278; the third year, .248. What was his overall average at that point? (Round to the nearest hundredth to estimate.)

   a. QUESTION:

   b. NECESSARY INFORMATION (*estimated*):

   c. NUMBER SENTENCE:

   d. ESTIMATE:

   e. CHECK (*go back and do the steps with exact figures*):

Answers are on page 113.

Comparing Decimal Numbers

To work with decimal problems, you need to **compare** the values of decimal numbers. For example, suppose you saw two popcorn stands at the zoo, one selling a bag of popcorn for $.90 and the other marked $.09 for the same size. At which stand would you buy your popcorn? You would go to the $.09 popcorn stand because you would rather pay 9 cents than 90 cents.

▶ Identify the better deal by circling the lower value in each case given below.

1. Pencils at $.40 or $.04 each

2. Candy bars at $.03 or $.30 each

3. Toothpaste at $1.87 or $1.78 a tube

4. A poster at $12.99 or $12.00

In finding the better buy, you saw that comparing the decimal values (after the decimal point) was just like comparing whole numbers.

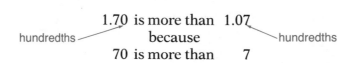

hundredths — 1.70 is more than 1.07 — hundredths
because
70 is more than 7

What if you need to figure out which is bigger, 3.5 inches or 3.05 inches, on a blueprint drawing? To compare these decimal numbers, add a zero to the end of 3.5, making it 3.50.

Now which is bigger, 3.05 or 3.50? You should be able to see that 3.50 is bigger, since 50 is bigger than 5. You are comparing 50 hundredths to 5 hundredths.

HINT: Decimals can be compared only if they have the same number of places. Put a zero or zeros at the end of the decimal that has the fewest number of places.

▶ Add enough zeros to give the decimals the same number of places. Then underline the largest in each group of decimals.

5. 7.7 yards or 7.07 yards

6. 16.09 feet or 16.9 feet

7. .03 inches or .3 inches or .003 inches

8. .6 ounces or .66 ounces or .069 ounces

Answers are on page 113.

Metric Measurement with Decimals

Metric measurements are based on decimals. In the United States we use metrics to measure liquids (2 liters of soft drink), the size of some foreign car parts, and medicine. Other countries use the metric system exclusively because the metric measurement system, based on decimals, is easy to work with.

Units of metric measurements are defined by their prefixes. Look at the chart below.

Prefix		Meaning	Example of Use
kilo	=	1,000	kilometer (1,000 meters)
hecto	=	100	hectoliter (100 liters)
deka	=	10	dekagram (10 grams)
deci	=	.1	decimeter (one-tenth of a meter)
centi	=	.01	centiliter (one-hundredth of a liter)
milli	=	.001	milligram (one-thousandth of a gram)

▶ Answer the following questions based on the chart above.

1. Which is largest?   kilometer   hectometer   dekameter

2. Which is smallest?   decigram   centigram   milligram

3. How many deciliters make 1 liter? _____

4. How many centimeters make 1 meter? _____

5. How many milligrams make 1 gram? _____

6. A liter is about the same size as a quart. Which of these would most likely be the measurement for a glass of milk?

   a. 250 milliliters   **b.** 2 liters   **c.** 1 kiloliter

7. A meter is a little larger than a yard. Which of these would most likely measure the width of a city lot?

   a. 5 kilometers   **b.** 25 meters   **c.** 250 millimeters

8. About 30 grams make one ounce. Which of these would most likely measure how much sodium (salt) is in a piece of white bread?

   a. 140 grams   **b.** 140 milligrams   **c.** 140 kilograms

Answers are on page 113.

Paychecks and Checking Accounts

Every time you examine the stub of your paycheck, spend money you have earned, or use a checking account, you are working with decimals. Look at the sample paycheck stub below.

Name: Ross, Lisa	Gross Pay	Federal Income Tax	State Income Tax	Social Security	Net Pay
Current Pay Period (*one week*)	$201.60	$25.87	$4.29	$16.08	$155.36
Year-to-Date	$8,064.00	$1,034.80	$171.60	$643.20	$6214.40

Lisa works 40 hours a week.

What is her gross pay per hour? $201.60 ÷ 40 = **$5.04**

What is her net pay per hour? $155.36 ÷ 40 = **$3.88**
(Lisa's $155.36 net pay divided by 40 hours equals $3.884 rounded to **$3.88**.)

▶ Complete the following statements, and answer the questions.

1. Gross pay is the pay that Lisa receives _____ taxes and social security are taken out.

2. Net pay is the pay that Lisa receives _____ taxes and social security are taken out.

3. Net pay is found by _____ the deductions from the gross pay.

4. The deductions taken from Lisa's pay are _____,
_____, and _____.

5. **a.** Based on Lisa's paycheck stub, what operation would you use to find how many weeks she has been working for this company?

_____

   b. Solve to find how many weeks that is. _____

6. a. What operations would you use to find how much Lisa would gross per week if she got a raise of 15 cents per hour? _____ and

_____.

b. What would her new gross weekly pay be? _____.

7. a. What operation would you use to find how much Lisa's deductions amount to each week? _____

b. Find the total of her weekly deductions. _____

Lisa deposited her pay into her checking account on August 8th.

NUMBER	DATE	TRANSACTION	PAYMENT/DEBIT	DEPOSIT/CREDIT	BALANCE
	8/1			$201.60	$657.82
1032	8/1	Electric Company	$15.59		642.23
1033	8/2	Fashion Fair	$36.66		605.57
1034	8/2	Iris's Grocery	$28.01		577.56
	8/8			$201.60	

8. What is Lisa's new balance on August 8th? _____
HINT: You *subtract* a payment or debit (a check you write). You *add* a deposit or credit.

▶ Now write in the following transactions that Lisa made and find her new balance.

9. 8/9   Check to Squeeky Cleaners for $12.00

10. 8/10   Check for $100.00 cash

11. 8/11   Check to phone company for $30.47

12. 8/15   Deposit weekly paycheck of $201.60 less $50.00 cash that she got back.

13. 8/15   Debit of $4.25 for new checks ordered.

Answers are on page 114.

Finding Total Cost

If you know the **unit price** of an item and how many units you are buying, you can figure the **total cost**. For example, at Bailey's Little League Ball Park hot dogs sell for $.88 each including tax. How much would four hot dogs cost?

> Total cost = unit price × number of units
> Total cost = $.88 × 4 hot dogs = **$3.52**

Unit prices also apply to items sold by measurement. For example, to find the total cost of 15 pounds of potatoes at $.22 a pound, the same formula is used.

> Total cost = $.22 (unit price) × 15 (number of units)
> Total cost = **$3.30**

▶ Find the total cost for these items:

1. 3 paperback books at $4.95 each = _____

2. 5 packages of gum at $.35 each = _____

3. 16 tomato plants at $1.09 each = _____

4. 7 yards of material at $6.45 a yard = _____

5. 3.2 pounds of meat at $2.79 a pound = _____

▶ Circle the letter of the closest estimated answer in each problem below.

6. Ed bought 4.7 pounds of nails that were priced 42 cents a pound. About how much did he spend on nails?

   a. $2.00   **b.** $3.00   **c.** $5.00   **d.** $20.00

7. Marcy is docked for tardiness. One morning she was 6 minutes late (.10 of an hour). At her hourly pay of $4.15 per hour, about how much did her tardiness cost her?

   a. $.20   **b.** $.30   **c.** $.40   **d.** $.50

8. A company cafeteria charges employees $.29 for every half-pint of juice they take from a dispenser. Joellen drinks a pint of juice. About how much is she charged?

   a. $.30   **b.** $.60   **c.** $.90   **d.** $1.20

Answers are on page 114.

Finding the Best Buy

We all want to get the best buy when we shop. To do so, we must be able to compare unit prices of items. Figure the **best buy** on pencils among these choices:

    a. package of 10 for $1.20
    b. package of 8 for $.89
    c. package of 3 for $.50

To find the best buy, we need to find the unit price, the price for one pencil in each package: Unit price = total cost ÷ number of units

    a. $\frac{\$1.20}{10} = \$.12$ each   **b.** $\frac{\$.89}{8} = \$.11$ each   **c.** $\frac{\$.50}{3} = \$.17$ each

So a package of **8 for $.89** is the best buy because we pay only $.11 per pencil.

The same formula would apply when finding the best buy on items sold in measured units. For example:

        3 lb tomatoes   5 lb tomatoes
            $1.99 $3.59

price per 1 lb = $\frac{\$1.99}{3\,lb}$ =   price per 1 lb = $\frac{\$3.59}{5\,lb}$ = $7.18
$.663 = $.66/lb $.718 = $.72 lb

You can see that **3 pounds at $1.99** is the better buy because you pay less per pound.

▶ Find the unit price and best buy in each of these situations.

NOTE: For some problems you may need to carry your answer to three decimal places instead of two.

1. a. roll of 12-exposure film at $2.25   **b.** roll of 24-exposure film at $2.99   **c.** roll of 36-exposure film at $3.69

2. a. pkg. of 50 letter-size envelopes at $1.10   **b.** pkg. of 75 letter-size envelopes at $.88   **c.** pkg. of 150 letter-size envelopes at $3.50

3. a. 6 hangers for $.70   **b.** 10 hangers for $1.49   **c.** 12 hangers for $1.59

4. a. 60 aspirin at $2.39   **b.** 100 aspirin at $2.99   **c.** 150 aspirin at $6.00

Answers are on page 114.

Multistep Problems

Decimal problems cannot always be solved by using only one step. Look at the newspaper ad below.

> **Apple Tree Apartments**
> $500/$610   1 bdrm/2 bdrms
> Security deposit—1 month's rent
> Senior citizen's discount—$50 off/mo

Gladys is a 68-year-old senior citizen who wants to rent a one-bedroom apartment at Apple Tree. How much would Gladys have to pay for rent and security deposit the first month?

Let's break this problem down into smaller problems.

QUESTION: How much would Gladys have to pay for rent and security deposit the first month?

Problem 1	**Problem 2**	**Problem 3**
What is 1 month's rent?	What is 1 month's security deposit?	What is the first month's total rent and security deposit?
$500 − $50 = $450	$500 − $50 = $450	$450 + $450 = **$900**

You could write the steps of the problem as:

$$(\$500 - \$50) + (\$500 - \$50) = \textbf{\$900}$$

Each set of parentheses stands for one step of the problem.

▶ Solve the following problems by making smaller ones out of them. Always check your final answer against the question to be sure that it makes sense.

1. A mother and her son are both getting braces on their teeth. The cost for the mother's braces will be $180 per month. The son's will cost $100 a month. How much will they pay after 24 months?

Problem 1	**Problem 2**

2. Alisha made 48 ounces of fruit punch by mixing 8-ounce cans of pineapple juice, orange juice, and cherry drink. How many cans of each does she need if she uses equal amounts of each kind?

Problem 1	**Problem 2**

3. A nutrition chart on a box of cereal says that a 4-ounce serving yields 110 calories, 1 gram of protein, and 25 milligrams of potassium. How many calories would the entire 24-ounce box supply?

Problem 1	**Problem 2**

What information in this problem is unnecessary? Why? _____
_____
_____
_____

4. The Bonzers currently rent videos from the View Video Store for $2 each. If they join the video club, they pay $25 to join, then $1 a movie up to 50 movies for the year. Assuming that either way they will rent 50 movies, how much money will they save by the end of a year if they join the club?

Problem 1	**Problem 2**
Problem 3	**Problem 4**

Answers are on page 114.

6 Fractions

You have worked with decimals, so you are familiar with parts of a whole. A fraction is another way of showing a part of a whole. *Fractions* are usually used when the whole has been divided into a number of parts. The fraction $\frac{1}{2}$, for example, shows one part of a whole that has two parts.

In this chapter you will work with:

- adding and subtracting fractions
- multiplying and dividing fractions
- ratios and proportions

Writing Fractions

In the section on decimals, you learned that the U.S. money system is based on decimals. Our coins (except silver dollars) represent fractions of a dollar. Look at the illustrations and fill in the blanks that make the statements true.

1 cent

one penny = $\frac{1}{100}$ of a dollar

_____ pennies make a dollar

5 cents

one nickel = $\frac{1}{20}$ of a dollar

_____ nickels make a dollar

10 cents

one dime = $\frac{1}{10}$ of a dollar

_____ dimes make a dollar

25 cents

one quarter = $\frac{1}{4}$ of a dollar

_____ quarters make a dollar

50 cents

one half-dollar = $\frac{1}{2}$ of a dollar

_____ half-dollars make a dollar

You should have filled in the blanks with **100, 20, 10, 4,** and **2** because 100 pennies make a dollar, 20 nickels make a dollar, 10 dimes make a dollar, 4 quarters make a dollar, and 2 half-dollars make a dollar.

▶ Identify what fraction of a dollar is shown in each of the following problems. The first is done for you.

1. 3 dimes = $\frac{3}{10}$ of a dollar   or _____ of a dollar

2. 2 pennies = _____ of a dollar   or _____ of a dollar

3. 3 quarters = _____ of a dollar   or _____ of a dollar

4. 1 half-dollar = _____ of a dollar   or _____ of a dollar

5. 4 nickels = _____ of a dollar   or _____ of a dollar

▶ For the following problems, write the correct fraction to stand for the part-to-whole relationship.

6. Kelly got 37 out of 50 questions correct. She got _____ of them right.

7. Joe bought 8 of 9 place mats that were left on the sale table. He bought _____ of the place mats.

8. Nita paid $2 for a vase that had a $3 price tag. She bought it for _____ of the price.

9. Suzanne lost 4 of the 15 pounds she wanted to lose. She achieved _____ of her goal.

10. Greg read 121 pages of a 330-page book in one sitting. He read _____ of the book in one sitting.

Answers are on page 115.

Adding and Subtracting with Common Denominators

You know that fractions represent a part of a whole. Look at the circle divided into eight parts. Notice that five sections are shaded.

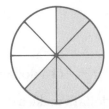

$$\frac{5}{8} \xleftarrow{\hspace{1cm}} \text{numerator}$$
$$\phantom{\frac{5}{8}} \xleftarrow{\hspace{0.5cm}} \text{denominator}$$

The top number, the ***numerator***, tells how many parts are in a particular fraction. The bottom number, the ***denominator***, tells the total number of parts. To add or subtract fractions, simply add or subtract the numerators. You can add or subtract fractions *only when the denominators are the same.*

$$\frac{2}{8} + \frac{3}{8} = \frac{5}{8} \qquad\qquad\qquad\qquad \frac{4}{7} - \frac{3}{7} = \frac{1}{7}$$

What do you do when there are different denominators? For example: $\frac{5}{8} + \frac{1}{2}$

Look at the larger denominator—8. Is it evenly divisible by the smaller denominator—2? In this case, yes, it is.

$$2)\overline{8}^{\,4}$$

Write both fractions with the same denominator. (You multiply by 4 to change the smaller fraction to higher terms.) Now you can add the numerators.

$$\begin{array}{r} \frac{5}{8} = \frac{5}{8} \\ + \frac{1}{2}\,{\scriptstyle(\times 4)} = \frac{4}{8} \\ \hline \frac{9}{8} = 1\frac{1}{8} \end{array}$$

What do you do when the smaller denominator won't divide evenly into the larger? For example: $\frac{3}{4} - \frac{2}{3}$

Multiply the denominators to get a common denominator.

$$\frac{2}{3} \quad\quad \frac{3}{4}$$
$$3 \times 4 = 12$$

Now, multiply both the numerators and denominators to give both fractions the same denominator.

$$\begin{array}{r} \frac{3}{4}\,{\scriptstyle(\times 3)} = \frac{9}{12} \\ - \frac{2}{3}\,{\scriptstyle(\times 4)} = \frac{8}{12} \\ \hline \frac{1}{12} \end{array}$$

Sometimes, when you multiply denominators, the common denominator will be very large. You may want to try to use a multiple of both numbers as a common denominator. For example:

$$\frac{5}{12} - \frac{3}{10}$$

If you multiplied, the denominator would be 120. Let's look at common multiples of both. The number 60 is a common multiple.

12: 12, 24, 36, 48, 60
10: 10, 20, 30, 40, 50, 60

Now, multiply both numerators and denominators so both have a common denominator of 60. These numbers are easier to work with.

$$\frac{5}{12} \frac{(\times 5)}{(\times 5)} = \frac{25}{60}$$
$$-\frac{3}{10} \frac{(\times 6)}{(\times 6)} = \frac{18}{60}$$
$$\frac{}{\quad} \quad \frac{7}{60}$$

Whatever number you multiply the denominator by, the numerator of that fraction must be multiplied by the same number.

Find the lowest common denominator for the following three fractions:

$$\frac{2}{3}, \frac{3}{4}, \frac{1}{6}$$

You're right if you said **12**. That's because 12 is a multiple of 3, 4, and 6.

3: 3, 6, 9, 12
4: 4, 8, 12
6: 6, 12

HINT: When adding and subtracting fractions, you must always find the *lowest* common denominator. This is the smallest number that results when each denominator is multiplied by another number.

▶ Find the lowest common denominator for each pair of fractions.

1. $\frac{2}{3}$ and $\frac{4}{8}$

2. $\frac{5}{6}$ and $\frac{7}{24}$

3. $\frac{1}{3}$ and $\frac{5}{7}$

4. $\frac{4}{9}$ and $\frac{5}{8}$

5. $\frac{4}{7}$ and $\frac{13}{21}$

6. $\frac{5}{10}$ and $\frac{13}{20}$

▶ Read the problems below. First find a common denominator in each problem. Then solve the problem.

7. $\frac{3}{7} + \frac{4}{21}$

8. $\frac{2}{3} + \frac{5}{13}$

9. $\frac{5}{8} - \frac{1}{2}$

10. $\frac{4}{5} - \frac{1}{20}$

Answers are on page 115.

Multiplying Fractions

Many word problems involving fractions are difficult to solve because it can be hard to determine which operation to use. This is especially true for problems that involve multiplication and division. You learned on page 16 that the word *of* often means to multiply. You can use the **"of" formula** to solve these types of problems. Look at the problem below.

Ginny broke $\frac{1}{3}$ of the dozen eggs she bought before she got home. How many did she break?

In this problem and in all multiplication problems involving fractions, the **whole** is given. You must solve to find the **part**. If you can translate the problem into an arithmetic expression, using the "of" formula, you can solve the problem.

QUESTION:	A fraction of	the whole	equals what part?
EXPRESSION:	$\frac{1}{3} \times$	a dozen eggs	= the number broken
SETUP:	$\frac{1}{3} \times$	12	= ?
SOLUTION:	$\frac{1}{3} \times$	12	= **4 eggs**

HINT: Whenever you are asked to find a *fraction of* something, you multiply.

▶ In the following problems, write the arithmetic expressions for the questions and then solve.

1. Theo and Marla watched $\frac{5}{6}$ of a long movie before going to bed. The movie was 204 minutes long. How much time did they spend watching it?

   QUESTION: A fraction of the whole equals what part?

   EXPRESSION:

   SETUP:

   SOLUTION:

2. Estimates are that $\frac{2}{5}$ of all new cars sold in the United States require repairs within the first few months of ownership. About 550,000 new cars were sold one year. According to the estimate, how many of them needed repairs within the first few months?

QUESTION: A fraction of the whole equals what part?

EXPRESSION:

SETUP:

SOLUTION:

3. Cecilia bought $\frac{3}{4}$ of the lace trimming that a small store stocked. The store had $5\frac{1}{3}$ yards before Cecilia arrived. How much lace did she buy?

QUESTION: A fraction of the whole equals what part?

EXPRESSION:

SETUP:

SOLUTION:

4. Tony ate $\frac{5}{8}$ of a box of chocolates. The full box of candy weighed $1\frac{1}{5}$ pounds. What was the weight of the chocolates that Tony ate?

QUESTION: A fraction of the whole equals what part?

EXPRESSION:

SETUP:

SOLUTION:

Answers are on page 115.

Dividing Fractions

Fraction problems involving multiplication and division are often difficult to distinguish because they look so similar. Many of these problems can be solved by applying the "of" formula.

In the previous lesson you saw that in a fraction problem in which the whole is given, the word *of* calls for you to multiply, as in this case:

Joan read $\frac{1}{2}$ of the 226-page book.

QUESTION: A fraction of   the whole   equals what part?

SOLUTION: $\frac{1}{2} \times$ 226   = 113 pages

In another type of word problem that includes the word *of*, the whole is not given. The **part** is given, and we need to find the **whole**. To solve this type of problem, we use an arithmetic expression also, but we *divide* instead of multiply.

Look at the following example.

Joan finished $\frac{1}{2}$ of her book in one night. She read 152 pages. How many pages long was the entire book?

QUESTION: A fraction of   the whole   equals what part?

EXPRESSION: $\frac{1}{2} \times$ entire book   = 152 pages

                   $\frac{1}{2} \times$ $\boxed{?}$ = 152

Since we don't know the whole, we must divide.

SOLUTION: $\frac{1}{2} \times \boxed{?} = 152$
$\boxed{?} = 152 \div \frac{1}{2} = 152 \times \frac{2}{1}$
$\boxed{?} = 304$

When you divide fractions, you **invert** (switch) the denominator and numerator and then multiply. Notice that when $\frac{1}{2}$ is inverted, it becomes $\frac{2}{1}$.

▶ In the following problems, write the expression for each question and then solve.

1. If the 48 factory employees who sign up for optional eye care insurance represent $\frac{2}{3}$ of the work force, how many people are employed by this factory?

QUESTION: A fraction of the whole equals what part?

EXPRESSION:

SETUP:

SOLUTION:

2. The Hillermons spent $13,440 on their new car. If this is $\frac{3}{5}$ of their combined yearly income, how much do they earn together in one year?

QUESTION: A fraction of the whole equals what part?

EXPRESSION:

SETUP:

SOLUTION:

3. Jeanette bought $\frac{7}{8}$ pound of bananas for $.56. How much would a full pound have cost her?

QUESTION: A fraction of the whole equals what part?

EXPRESSION:

SETUP:

SOLUTION:

4. A cereal box indicates that $\frac{7}{8}$ of a cup of cereal contains 105 calories. How many calories does a full cup of the cereal contain?

QUESTION: A fraction of the whole equals what part?

EXPRESSION:

SETUP:

SOLUTION:

Answers are on page 115.

Deciding Whether to Multiply or Divide

Now that you have practiced solving multiplication and division problems that involve fractions, you are ready to practice deciding when a problem calls for one or the other operation. Look at the example below.

Sam ran $6\frac{1}{2}$ miles in one week. This was only $\frac{2}{3}$ of what he had planned to run. How far had he planned to run?

QUESTION: A fraction of   the whole   equals what part?

$\frac{2}{3} \times$ distance planned $= 6\frac{1}{2}$

OPERATION: multiply or (divide)? (*circle one*)

REASON: part not given / (whole not given) (*circle one*)

EXPRESSION: $\frac{2}{3} \times \boxed{?} = 6\frac{1}{2}$

SOLUTION: $\frac{2}{3} \times \boxed{?} = 6\frac{1}{2}$

$\boxed{?} = 6\frac{1}{2} \div \frac{2}{3}$

$\boxed{?} = 6\frac{1}{2} \times \frac{3}{2}$

$\boxed{?} = \frac{13}{2} \times \frac{3}{2} = \frac{39}{4} = \textbf{9}\frac{\textbf{3}}{\textbf{4}} \textbf{ miles}$

▶ Solve the following problems by using the "of" formula. As in the example, first decide whether the problem requires you to find the whole (divide) or find the part (multiply). Then tell why you chose that operation. When dividing fractions, be sure to invert the fraction before multiplying.

1. Grace used $\frac{2}{3}$ of her supply of flour to bake cookies. Her supply was $12\frac{1}{2}$ cups of flour. How much did she use?

    QUESTION: A fraction of the whole equals what part?

    EXPRESSION:

    OPERATION: multiply or divide? (*circle one*)

    REASON: part not given / whole not given (*circle one*)

    SOLUTION:

2. Phil knows that when his car has $\frac{4}{5}$ of a tank of gas, he has 18.4 gallons of gas left. How many gallons does his tank hold?

QUESTION: A fraction of the whole equals what part?

EXPRESSION:

OPERATION: multiply or divide? (*circle one*)

REASON: part not given / whole not given (*circle one*)

SOLUTION:

3. Tracy got $\frac{5}{10}$ of the questions wrong on her test. If she missed 25 questions, how many questions were on the test?

QUESTION: A fraction of the whole equals what part?

EXPRESSION:

OPERATION: multiply or divide? (*circle one*)

REASON: part not given / whole not given (*circle one*)

SOLUTION:

4. The Bruskas made a down payment of $\frac{2}{7}$ of the purchase price on a new television set. The TV's sale price was $357.35. How much was their down payment?

QUESTION: A fraction of the whole equals what part?

EXPRESSION:

OPERATION: multiply or divide? (*circle one*)

REASON: part not given / whole not given (*circle one*)

SOLUTION:

Answers are on pages 115–16.

Ratios and Proportions

A *ratio* is a kind of fraction. For example, we can say that $\frac{3}{4}$ of all doctors surveyed recommend using "Pain Away." This statement can be written as *3 out of 4*, or $\frac{3}{4}$, or as a ratio *3:4*.

▶ Write each of the following expressions as a ratio and as a fraction.

1. 7 out of 8 dentists

fraction _____ ratio _____

2. 9 out of 10 mail carriers

fraction _____ ratio _____

3. decade : century

fraction _____ ratio _____

4. foot : feet in a yard

fraction _____ ratio _____

A *proportion* is a pair of equal fractions. Proportions are used to show that two relationships are equal. For instance, if 6 out of 8 doctors recommend "Pain Away," this is really the same proportion as 3 out of 4. Why? Because $\frac{6}{8}$ can be reduced to $\frac{3}{4}$; they are *equivalent fractions*.

As a proportion, this is written as the pair of equivalent fractions:

$$\frac{3}{4} = \frac{6}{8}$$

This is read as *3 is to 4 as 6 is to 8*. Another way of writing it is as an *analogy*: 3:4 :: 6:8.

▶ Look at the following proportions written as analogies. Fill in the missing pictures or words.

5. ☐ : ⊞ :: _____ : ⊕

6. ▯ : ▢ :: ◺ : _____

7. gross pay : net pay :: original price : _____

    a. tax deductions **b.** discount rate **c.** sale price **d.** list price

8. ounce : pound :: _____ : 16

    a. $\frac{1}{16}$ **b.** 1 **c.** 16 **d.** 160

9. inches : foot :: _____ : pound

    a. yards **b.** ounces **c.** tons **d.** heavy

Answers are on page 116.

Setting Up Proportions

Some problems are best solved by setting up a proportion that has a missing piece. This missing piece is the solution to the problem. Read the problem below.

If Nora can travel 100 miles on 5 gallons of gas, how much gas would she need to go 220 miles?

The correct setup for this proportion would be:

$$\frac{100 \text{ miles}}{5 \text{ gallons}} = \frac{220 \text{ miles}}{x \text{ gallons}}$$

The first part of the proportion is given: 100 miles to 5 gallons of gas, or $\frac{100}{5}$. For the second part of the proportion, the equivalent miles are given (220), so we must solve for the equivalent gallons.

HINT: It's good to get used to writing x for an unknown number. In the problem above, you would write x gallons.

▶ For each problem below, just set up the proportion. Use x for the unknown. *Do not solve.*

1. If Tim receives $144 for 12 hours of work, what would he receive at the same rate of pay for 40 hours of work?

2. If a school has 10 teachers for every 180 students, how many teachers would it need in order to keep the same ratio for 216 students?

3. Mrs. Fendella's math class can do 30 problems in 10 minutes. At that rate, how many minutes would it take them to do 21 problems?

4. Jack spent $6.69 for 3 pounds of candy. How much would he spend for 2.5 pounds of the same kind of candy?

5. The shadow of the tree to the left is 8 feet. What is the length of the shadow of the other tree?

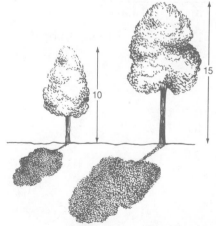

Answers are on page 116.

Solving Proportions

In a proportion, the cross-products are equal. **Cross-products** are the products of the numbers at opposite corners of a proportion.

$$\frac{7}{8} \times \frac{14}{16} \qquad \begin{array}{l} 7 \times 16 = 112 \\ 8 \times 14 = 112 \end{array} \qquad \text{Cross-products are equal.}$$

Often, you will need to find a missing number in a proportion. Letters are used to stand for unknown numbers. To find the missing number in a proportion cross-multiply.

Set up the proportion.

$$\frac{5}{6} = \frac{x}{30}$$

Cross-multiply and divide.
$$6x = 150$$
$$6x \div 6 = 150 \div 6$$
$$x = 25$$

The proportion is $\frac{5}{6} = \frac{25}{30}$, which is read as "5 is to 6 as 25 is to 30."

▶ Solve the following proportions for x.

1. $\frac{1}{7} = \frac{3}{x}$

   a. 10   **b.** 14   **c.** 21   **d.** 28

2. $\frac{10 \text{ men}}{40 \text{ women}} = \frac{15 \text{ men}}{x}$

   a. 20 women   **b.** 45 women   **c.** 60 women   **d.** 150 women

3. $\frac{12 \text{ birds}}{3 \text{ cages}} = \frac{x}{4 \text{ cages}}$

   a. 13 birds   **b.** 16 birds   **c.** 18 birds   **d.** 48 birds

4. $\frac{\$10}{15 \text{ lb}} = \frac{\$30}{x}$

   a. 5 lb   **b.** 20 lb   **c.** 35 lb   **d.** 45 lb

Answers are on page 116.

Solving Word Problems Using Proportions

Now let's put together setting up proportions and solving them. If you're able to see problem solving with proportions as a two-step process, it will be easier for you to solve word problems using proportions. Look at the following example.

Pete's Pet Shop keeps a maximum of 10 birds for every 3 cages it owns. If Pete just expanded his store to hold 15 cages, what is the maximum number of birds he can keep?

Set up the proportion.   Cross-multiply and divide.

$$\frac{10 \text{ birds}}{3 \text{ cages}} = \frac{x \text{ birds}}{15 \text{ cages}}$$

$3x = 150$

$x = 150 \div 3 = \textbf{50 birds}$

HINT: The most important step is to set up the proportion correctly. The known relationship should be the first ratio—such as "10 birds to 3 cages." The second ratio must follow the same pattern—"birds to cages."

▶ Now set up the proportions and solve the following problems.

1. If Maria can set up 2 sets of files in 10 working hours, how long will it take her to set up 25 sets of files?

2. The owners of the Sea-Worthy Restaurant plan to cover several posts with rope for decoration. They need 12 feet of rope for each foot of post. How much rope will they need for each 7-foot-high post?

3. For every 13 students, a night school class orders 39 books. How many books would the school have to order for an enrollment of 156 students?

4. A recipe for homemade syrup calls for 3 cups of sugar for every 2 cups of water. How much water would be needed with 9 cups of sugar?

Answers are on page 116.

7 Percents

You see percents used in some form every day. Percents are used to represent discounts on sale items, tax rates, interest rates, inflation rates, and unemployment rates.

To understand percents, you must understand decimals and fractions. Decimals, fractions, and percents are all closely related and are different methods of expressing the same number relationship.

In this chapter you will learn:

- using fractions, decimals, and percents
- finding the part, the whole, and the percent
- solving for interest

When to Use Decimals, Fractions, and Percents

Decimals, fractions, and percents all relate parts to a whole. However, each has specific uses and advantages. Look at the example below to see how each expression represents the relationship of a quarter to a dollar.

    .25 (decimal) $\frac{25}{100}$ or $\frac{1}{4}$ (fraction)   25% (percent)

▶ Write the letter that indicates whether each of the following is best expressed using a fraction (f), a decimal (d), or a percent (p).

_____ 1. showing a batting average

_____ 2. revealing a rate of interest on a loan

_____ 3. telling an amount of money in U.S. dollars and cents

_____ 4. identifying an exact metric measurement

_____ 5. describing less than one inch on a ruler

_____ 6. advertising a rate of discount in a sale

Answers are on page 116.

Percents as Ratios and Fractions

You learned earlier that a ratio or fraction compares two numbers. Like a fraction, a percent also shows a part/whole relationship. For example, 50% can be expressed as this fraction:

$$\frac{50}{100}$$ the number to the left of the percent sign
always 100

The bottom part of the ratio is always 100 because percent means *of every 100*.

A number sentence expressed as a percent might be written:

50% of the population is female = 50 people out of every 100
people are female

HINT: While you must reduce fractions to their lowest terms, you do not reduce ratios.

▶ Fill in the blanks to make the sentences true and the ratios equivalent. The first is done for you as an example.

Ratio

1. Of the shipment, 72% arrived = __72__ out of every __100__ parts of the shipment arrived. $\frac{72}{100}$

2. Of the town's population, 9% is over 80 years old = _____ out of every _____ people are over age 80.   _____

3. Of the scores, 17% are below average = _____ out of every _____ scores are below average.   _____

4. Of the graduating students, 98% went to college = _____ out of every _____ students went to college.   _____

5. Of the houses, $33\frac{1}{3}$% were destroyed by the fire = _____ out of every _____ houses were destroyed.   _____

6. A 6.5% sales tax was added to the cost = _____ cents were added to every _____ cents of the cost.   _____

7. Cost of living has increased 200% from what it was a decade ago = it takes _____ dollars today to buy an item that cost _____ ten years ago.   _____

Answers are on page 117.

Percents as Decimals

Percents can be expressed as ratios (fractions) and also as decimals, since all describe parts of a whole. Remember, 50% means 50 out of 100. Hundredths are indicated in decimals by two places to the left of the decimal point:

50% = .50. = .50   4% = .04. = .04   10.3% = .10.3 = .103

Notice that in whole numbers like 50% or 4%, no decimal point is written, but it is understood to be located at the right of the whole number. Also, notice that for percentages less than 10%, you have to add a zero so you can move the decimal two places to the left.

HINT: To change any percent to a decimal, simply drop the % sign and move the decimal point two places to the left.

▶ Change the following percents to decimals. The first is done for you as an example.

1. 1% of the group = __1__ of the group of 100. This is read as *one hundredth*.

2. 25% discount = _____ off each dollar

3. 8% sales tax = _____ tax rate

4. 12.9% interest rate = _____ rate

5. 110% of last year's crop = _____ times last year's crop

6. 300% markup = _____ times the price

To change decimals to percents, move the decimal point two places to the right and add a percent symbol (%):

.15 = .15. = 15%   .01 = .01. = 1%   .015 = .01.5 = 1.5%

HINT: Notice that the zeros in .01 and .015 are dropped when you changed these decimals to percents.

▶ Change these decimals to percents. The first is done for you as an example.

7. .076 sales tax = _7.6%_ tax rate

8. $.10 on the dollar discount = _____ discount rate

9. 2.00 times the original = _____ of the original

10. .311 batting average = the batter hits _____ of the pitches

Answers are on page 117.

Using the "Of" Formula with Percents

Remember the "of" formula we used with fractions? The same formula can be applied to percents.

Let's try translating some problems using the "of" formula. Look at the problem below.

> Wanda, a waitress, gets a tip of 15% of the bill before tax. If the Labelles' bill came to $28.33 *before* tax, how much of a tip did they leave for Wanda?

QUESTION: A percent of the whole equals what part?

EXPRESSION: 15% × $28.33   = ?

SOLUTION: .15 × $28.33   = $4.25

To find the part, just multiply.

▶ In the following problems, write the arithmetic expression for the questions and then solve.

1. Illinois has a 6.5% sales tax. If Beverly's purchases totaled $132.76, how much sales tax did she have to pay?

   QUESTION: A percent of the whole equals what part?

   EXPRESSION:

   SOLUTION:

2. Only 40% of the 3,905 registered voters in one town cast ballots in the last election. How many voted?

   QUESTION: A percent of the whole equals what part?

   EXPRESSION:

   SOLUTION:

3. Juan must get 75% correct on all his tests to earn a C. If the number of questions totals 300, how many will he have to get right to earn a C?

QUESTION: A percent of the whole equals what part?

EXPRESSION:

SOLUTION:

Now let's look at some problems that follow a pattern in which the whole is not given. Look at the example below.

Government workers were given a 3.2% raise. This gave Sung Won $38.40 more per month. How much was her monthly salary before the raise?

QUESTION: A percent of the whole equals what part?

EXPRESSION: 3.2% × [?] = $38.40 raise

SOLUTION: [?] = $38.40 ÷ .032

[?] = **$1,200**

Since the unknown whole (her salary) is part of the multiplication equation, you must divide to find it.

▶ In the following problems, write the arithmetic expressions for the questions and then solve.

4. Minneapolis got 12% more snow this year than last. The city got 10.08 more inches this year. How much snow did Minneapolis get last year?

QUESTION: A percent of the whole equals what part?

EXPRESSION:

SOLUTION:

5. Dinette sets were advertised at a store for 30% off the regular price. The Colonnas will save $300 on the set they have been eyeing. How much is the regular price?

QUESTION: A percent of the whole equals what part?

EXPRESSION:

SOLUTION:

6. If home prices are 150% of what they were two years ago and the price of one house is now $120,000, what was its price two years ago?

QUESTION: A percent of the whole equals what part?

EXPRESSION:

SOLUTION:

Now let's look at a third and final pattern in which the percent itself is not given. Look at the example below.

A taxi driver is given a $.60 tip on a $4.00 fare. What was the percent of the tip?

QUESTION: A percent of the whole equals what part?

EXPRESSION: $?$ × $4   = $.60

SOLUTION: $?$ = $.60 ÷ 4 = .15 = **15%**

Since the unknown is part of the multiplication expression, you divide.

▶ In the following problems, write the arithmetic expressions for the questions and then solve.

7. If a family's combined income is $585 a week and it spends $100 a week on food, what percent of its income is spent on food?

QUESTION: A percent of the whole equals what part?

EXPRESSION:

SOLUTION:

8. A store advertises a sale of $7 off on all men's shirts. What is the percent of savings on a $21 shirt at that store?

QUESTION: A percent of the whole equals what part?

EXPRESSION:

SOLUTION:

Answers are on page 117.

Using the Percent Circle

Now that you have practiced finding the part, the whole, and the percent with the "of" formula, try mixing them up. To make it easier to decide whether to multiply (to find the part) or divide (to find the whole or the percent), you can use a helpful tool called the percent circle:

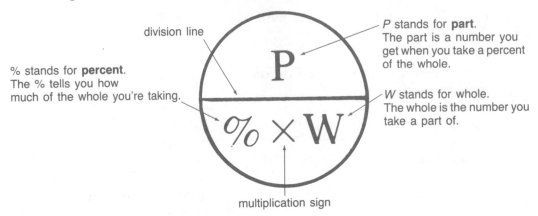

division line

% stands for **percent**.
The % tells you how
much of the whole you're taking.

P stands for **part**.
The part is a number you
get when you take a percent
of the whole.

W stands for whole.
The whole is the number you
take a part of.

multiplication sign

These three problems show you how you can use the percent circle to find a percent, part, or whole.

1. To find a **part** of the whole:
   If 20% of the people in a city are Hispanic and there are 200,000 people in the city, how many of them are Hispanic?

Cover the P (part), the number you are trying to find. That leaves you with a multiplication problem:
$P = \% \times W$
$P = 20\% \times 200,000$
$P = .20 \times 200,000 = \textbf{40,000 people}$

2. To find the **percent**:
   Mr. Park was told that he needed a down payment of $2,100 to buy his $14,000 car. What percent did he have to put down?

Cover the % (percent), the number you are trying to find. That leaves you with a division problem:
$\% = P \div W$
$\% = \dfrac{\$2,100}{\$14,000} = \dfrac{3}{20} = \textbf{15\%}$

3. To find the **whole**:

Tuition at Scott Community College went up 20%, for an increase of $10 per credit hour. What was the tuition previously?

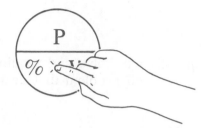

Cover the W (whole), the number you are trying to find. That leaves you with a division problem:

$$W = P \div \%$$

$$W = \frac{\$10}{.20} = \textbf{\$50 per credit hour}$$

HINT: Don't be confused that your answer, $50, is more than the $10 given in the problem. The $10 is just the part (how much the tuition increased) from the old tuition.

▶ Now use the percent circle to solve each of the problems below. First decide whether you are looking for the percent, the part, or the whole.

1. The population of Hometown is now 8,000 people. In 1940 it was 4,000 people. What percent of today's population was the 1940 population?

2. Mrs. Murray's annual salary of $19,000 will soon increase by 8.5%. By how many dollars will her salary increase?

3. Of the shipment of 1,100 stalks of bananas, 15% was damaged en route. How many stalks were damaged?

4. If the Rogers family paid $107.10 in taxes on a used car and the tax rate is 7%, what did the family pay for the car?

5. The owner of a pawnshop sold a bracelet for 350% of what he paid for it. If he sold it for $182, how much did he pay for it?

6. A family drove all day but covered only 25% of its journey. They had traveled 475 miles that day. How long will their journey be?

7. Adrian had to pay $6.75 sales tax on his purchase of $125. What tax rate did he have to pay?

Answers are on page 117.

Percent of Increase or Decrease

Some percent problems take more than one operation to solve and require some knowledge of key words. Let's try solving one such problem that deals with percents of increase.

A model of car that cost $9,000 in 1980 costs $12,000 now. What has been the percent of increase in the price?

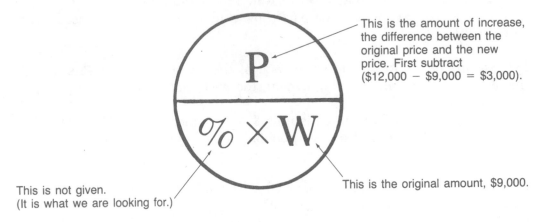

This is the amount of increase, the difference between the original price and the new price. First subtract ($12,000 − $9,000 = $3,000).

This is not given. (It is what we are looking for.)

This is the original amount, $9,000.

This percent-of-increase problem can be represented as:

$$\frac{\text{difference}}{\text{original amount}} = \frac{\$12,000 - \$9,000}{\$9,000}$$

$$\text{percent of increase} = \frac{3,000}{9,000} = \frac{1}{3} = 33\frac{1}{3}\%$$

▶ Refer to the percent circle to solve the following problems. Be sure to subtract first to find the amount of increase.

1. Inflation has caused the price of yams to rise from $.20 a pound in season in 1980 to $.30 a pound in season now. What has been the percent of growth in the price?

2. Last year Lawrence earned $6 an hour. Now he earns $7 an hour. What was the percent increase of his raise?

3. Belinda and Jack buy old houses, fix them up, and resell them at a profit. They put $50,000 into one and sold it for $90,000 several months later. What was their percent of profit?

4. The owners of Frannie's Frocks pay an average of $40 for a dress that they can sell for $120. What is their percent of markup?

5. Notice that percent-of-increase problems can also be identified by such words as _____, _____, _____, and _____.

Let's try some percent-of-decrease problems similar to the problems we have just completed.

> A typical community college class begins with 30 students but has only 25 by midterm. What is the percent of decrease in enrollment?

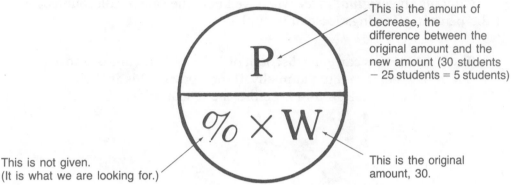

This is the amount of decrease, the difference between the original amount and the new amount (30 students − 25 students = 5 students)

This is not given. (It is what we are looking for.)

This is the original amount, 30.

Percent of decrease can also be written as:

$$\frac{\text{difference}}{\text{original amount}} = \frac{30 - 25}{30}$$

$$\text{percent of decrease} = \frac{5}{30} = \frac{1}{6} = \mathbf{16\frac{2}{3}\%}$$

▶ Solve the following percent-of-decrease problems.

6. Hal bought a pickup truck for $10,000 one year. It depreciated, and he was able to sell it for $8,000 the next year. What was the percent of depreciation?

7. Nicole works in a store that sells washing machines for $550. As an employee, she was able to buy one for only $495. What is her percent of discount?

8. A fire destroyed all but 3,000 of 12,000 two-by-fours at a lumberyard. What percent of the two-by-fours was lost in the fire?

9. A ham weighed 8 pounds before cooking, but weighed only 7 pounds after cooking. What was its percent of shrinkage?

10. Notice that percent-of-decrease problems can also be identified by such words as _____, _____, _____, and _____.

Answers are on page 117.

Original and Sale Price

You have learned that solving some percent problems requires more than one step or operation. Now you will work with one of every shopper's favorite subjects—sales.

Let's try finding the **_original price_** when we know the sale (or discounted) price and the percent of savings (discount rate).

> The Washingtons can get a discount of 15% on their car insurance if they pay their premium annually. If their payment is $637.50, what is the annual premium before the discount?

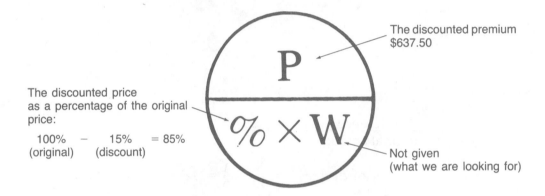

The discounted premium
$637.50

The discounted price as a percentage of the original price:

100% − 15% = 85%
(original) (discount)

Not given
(what we are looking for)

Original price can be found by:

$$\frac{\text{sale price}}{\text{percent of original price}} = \frac{\$637.50}{100\% - 15\%} = \frac{\$637.50}{.85} = \mathbf{\$750}$$

▶ Find the original price in the problems below.

1. Mary Ann paid $60 for a dress marked 25% off the original price. What was the original price?

2. A gas station gives a 5% discount for a cash purchase. If Rob paid $10.45 cash for his gas, what would the charge have been for a credit card purchase?

3. During the August White Sale, Kemp bought sheets at 30% off the regular price. He paid $24 for them. How much did they cost originally?

4. Lila went to an end-of-the-season clearance sale at which she purchased a coat for 75% off the original price. She paid $200 for it. What was the original price?

Now let's practice finding the *sale price* when we know the percent of savings (discount rate) and the original price.

Jenning gets a 10% employee discount on items sold in Harold's Hardware Store. He buys a lawn mower that regularly sells for $230. How much does he have to pay for it?

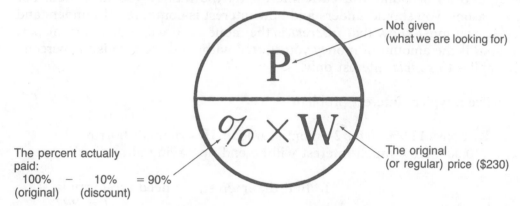

Not given
(what we are looking for)

The original
(or regular) price ($230)

The percent actually paid:
100% − 10% = 90%
(original) (discount)

Sale price can be found by:

original price − discount = $230 − (.10 × $230) = $230 − 23 = **$207**

▶ Find the sale price in the problems below.

5. Wesley convinced a car salesman to lower the price of a new car by 20% from the sticker price of $12,550. What did Wesley pay for the car?

6. Curtains were advertised at 40% off the original price. If Sally bought a pair that originally sold for $105, what did she pay for them?

7. An amusement park sells family season passes at a discount of $12\frac{1}{2}\%$. If the family passes regularly sell for $120, what is the discounted price for the passes?

8. A shoe store has a storewide sale featuring 30% off any shoes in stock. Anthony wants a pair of running shoes that normally sells for $50. How much would he pay for the shoes during this sale?

Answers are on page 117.

The Interest Formula

When you make credit purchases, you pay interest. Also, when you deposit money into an account, you earn interest on the money you deposited. For these reasons, you should understand how interest is computed. To understand interest, you must know that **interest** is the money you will either earn or pay, **principal** is the amount of money you started with, and the **rate** is the percent. This applies to *simple* interest only.

Let's solve a typical interest problem:

>   Kyle pays 12.9% on his 24-month car loan. He originally borrowed $10,000. How much interest will he end up paying altogether?

To find the interest, we need the formula
$$I = PRT$$
Interest = Principal × Rate × Time (in years)
$$I = \$10,000 \times 12.9\% \times 2$$
$$I = \$10,000 \times .129 \times 2 = \mathbf{\$2,580} \text{ interest}$$

▶ Figure the interest for the problems below. Use a calculator if you wish.

1. Morgana has a savings account that pays 5.75% interest. How much interest will she get after leaving $1,000 in her account for three years?
$$I = PRT$$

2. Rodney deposited $500 in his savings account a year ago. How much interest has he earned, considering his bank pays 6.5% interest?
$$I = PRT$$

3. Tammi borrowed $1,800 from her father and agreed to pay him back at 5% interest. How much interest did she owe him at the end of $2\frac{1}{2}$ years? HINT: Change $2\frac{1}{2}$ years to decimal form.
$$I = PRT$$

4. For her eighth-grade graduation, Sylvia received $835 in gifts. She used that money to open a savings account that pays 7.5% interest. If she doesn't make any more deposits, how much interest does she accumulate by the end of her freshman year in high school?
$$I = PRT$$

Sometimes you want to know the total you will have to **pay back** on a loan. At other times you may want to determine how much money you will have in a savings account by the end of a certain time period. To find these amounts, you have to add the interest you have earned to the original principal.

Let's figure the total for the example about Kyle on the facing page. We already found that the interest is $2,580. The principal is $10,000. So $10,000 + $2,580 = **$12,580**, the total paid on the loan.

$10,000 principal
+ 2,580 interest
$12,580 total repaid on the loan

HINT: Always read interest problems carefully. Make sure that you understand whether you are looking for the interest or the total to pay back.

▶ Now find the total paid in each of the problems you just solved.

5. Problem 1

   Interest = _____

   Principal = _____

   Total in account = _____

6. Problem 2

   Interest = _____

   Principal = _____

   Total in account = _____

7. Problem 3

   Interest = _____

   Principal = _____

   Total paid on loan = _____

8. Problem 4

   Interest = _____

   Principal = _____

   Total in account = _____

Answers are on page 118.

Answer Key

CHAPTER 1: NUMBER RELATIONSHIPS

The Multiplication Grid
pages 4–5

1. 2, 4, 6, 8, or 0

2. **a.** 15, 1 + 5 = 6
   b. 18, 1 + 8 = 9
   c. 21, 2 + 1 = 3
   d. 24, 2 + 4 = 6
   e. 27, 2 + 7 = 9
   f. 30, 3 + 0 = 3
   g. 3, 6, or 9

3. 5, 0

4. **a.** even
   b. 3

5. **a.** 9
   b. 2, 7
      3, 6
      4, 5
      5, 4
      6, 3
      7, 2
      8, 1
   c. 1, 1

6. 0

7. 40; 766

8. 55; 400

9. 70; 1,020

10. 90

Prime Numbers
pages 6–7

9. 2, 3, 5, 7, 11, 13, 17, 19, 23, 29, 31, 37, 41, 43, 47, 53, 59, 61, 67, 71, 73, 79 83, 89, 97

10. 25

Finding the Pattern
page 8

1. add the numbers
2. divide the first number by the second
3. add 8
4. subtract 1

Patterns and Series
page 9

1. **U** — The pattern is to supply the vowels.

2. **H** — The pattern is to follow the alphabet.

3. **V** — The pattern is to list consecutive letters of the alphabet in increases of 1. For example, R is listed once, S twice, T three times, and so on.

4. **M** — The pattern is to skip a letter of the alphabet.

5. **E, F** — The pattern is to list a pair of letters in order starting from the beginning of the alphabet and then a pair of letters in reverse order starting from the end of the alphabet.

6. **best**

7. **go**

8. **female**

9. **forward**

10. **November** — The pattern is to list the months of the year that have 30 days.

Measurement and Patterns
page 10

1. second, time, decreasing
2. dime, money, decreasing
3. mile, distance (length), increasing
4. gallon, liquid volume, decreasing

Numbers in a Series
page 11

1. **30** — to add 5
2. **73** — to subtract 9
3. **1** — to divide by 10
4. **65,536** — to multiply the number by itself
5. **778** — to multiply by 10 then subtract 2

6. 28 to add a number that is one greater than the last number added.

7. 16 to divide by 2 then add 2

8. 612 to subtract 6 then add 12

Number Patterns
page 12

1. 30, 35, 50 multiply

2. 26, 30, 108 subtract

3. 39, 54, 57, 59 $x + 6 = y$

4. 4, 5, 6, 7 $x \div 9 = y$

Estimating
page 13

1. $100

2. 10

3. $100

4. 10,000

5. 100,000

Rounding Numbers
pages 14–15

1. 250 pages

2. 6,700 people

3. 74,000 boxes

4. 200,000 births

5. $9,500,000

6. a. 1,206
b. 1,200

7. a. 963
b. 1,000

8. a. 264,152
b. 280,000

9. a. 20r21
b. 20

10. 500 + 700 + 500 = **1,250**

11. 26,700 − 9,500 = **17,200**

12. 900 ÷ 300 = **3**

13. 300 × 900 = **270,000**

14. b $5,000 × 70 = $350,000

15. b 6,000 ÷ 300 = 20

16. b 80 + 30 + 20 = 130

CHAPTER 2: FROM WORDS TO NUMBERS

Writing Expressions
pages 16–17

1. is   =

2. for, is, per   ÷, =

3. for, is, per   ÷, =

4. of, is   ×, =

5. of   ×, =

6. $160 ÷ 4

7. 7 × 200 lb

8. 10 × 24 crayons

9. 28 students ÷ 2

Equations and Unknowns
pages 18–19

1. 50 cents for 2 kiwi fruit

50 cents ÷ 2 = a
a = **25 cents**

$$
\begin{array}{r}
25 \\
2\overline{)50} \\
4 \\
\hline
10 \\
10 \\
\hline
\end{array}
$$

2. 6 of the $5 hand towels

6 × $5 = a
a = **$30**

$$
\begin{array}{r}
\$5 \\
\times\ 6 \\
\hline
\$30 \\
\end{array}
$$

3. 11 of the 60-watt light bulbs

11 × 60 watts = a
a = **660 watts**

$$
\begin{array}{r}
60 \\
\times\ 11 \\
\hline
60 \\
60 \\
\hline
660 \\
\end{array}
$$

4. $324 for 3 persons

$324 ÷ 3 = a
a = **$108**

$$
\begin{array}{r}
108 \\
3\overline{)324} \\
3 \\
\hline
2 \\
0 \\
\hline
24 \\
24 \\
\hline
\end{array}
$$

5. 12 of the 52-pound bales

52 × 12 = a
a = **624 pounds**

$$
\begin{array}{r}
52 \\
\times\ 12 \\
\hline
104 \\
52 \\
\hline
624 \\
\end{array}
$$

Choosing the Operation
pages 20–22

1. *blank*

2. +

3. ×

4. +

5. *blank*

6. *blank*

7. ÷

8. ÷

9. ÷

10. −

The Averaging Formula
page 23

1.

$$
\begin{array}{r}
79 \\
82 \\
80 \\
84 \\
+\ 75 \\
\hline
400 \\
\end{array}
\qquad
\begin{array}{r}
80 \\
5\overline{)400} \\
40 \\
\hline
0 \\
0 \\
\end{array}
$$

2.
```
   15        20 people
   21      6) 120
   23         12
   27          0
   19          0
 + 15
  120
```

3.
```
  $105        $102
    92      4) 408
   110         4
 + 101         0
  $408         0
               8
               8
```

4.
```
  4,050      3,026 miles
  2,106    3) 9,078
 + 2,922      9
  9,078       0
              0
              7
              6
             18
             18
```

5.
```
   17        28 years
   23      7) 196
   24        14
   26        56
   32        56
   34
 + 40
  196
```

The Distance Formula
pages 24–25

1.
```
   350
 ×   4
 1,400      D = 1,400 mi
```

2.
```
   35
 ×  3
  105       D = 105 mi
```

3.
```
     4 hrs     T = 4 hrs
 40) 160
     160
```

4.
```
    12 hrs     T = 12 hrs
 35) 420
    35
    70
    70
```

5.
```
       55 mph    R = 55 mph
 20) 1,100
     1 00
      100
      100
```

6.
```
      651 mph    R = 651 mph
 3) 1,953
    1 8
     15
     15
      3
      3
```

CHAPTER 3: PROBLEM-SOLVING STRATEGIES

Restating the Problem
pages 26–27

1. 55, 105, more, factory, end

2. five, quarts, themselves, how many quarts, split, evenly

3. Answers will vary.

4. Answers will vary.

5. c

Working Step by Step
pages 28–29

1.
```
   89,977
 + 57,886
  147,863
```

4.
```
        2,030 r9
 26) 52,789
     52
     78
     78
      9
      0
      9
```

2.
```
     376
  × 703
   1 128
   0 00
 263 2
 264,328
```

3.
```
  24,598
 − 19,872
   4,726
```

5. $12 \div 3 = 4 + 5 = 9$

6. $10 + 10 = 20 − 20 = 0$

The Five-Step Model
pages 30–31

1. question

2. information

3. operation

4. answer

5. logical

6. (1) How many two-inch-high frozen dinners can be stacked into a freezer 36 inches high?

(2) 36 inches high, 2 inches high

(3) divide

(4)
```
     18 dinners
  2) 36
```

(5) It is logical that a freezer can hold 18 two-inch-high frozen dinners.

7. (1) How much will the contestant win after 50 seconds?

(2) $1 for first 10 seconds, doubles every 10 seconds, 50 seconds total time

(3) multiply and add

(4) $1 (*10 sec*) + $2 (*20 sec*) + $4 (*30 sec*) + $8 (*40 sec*) + $16 (*50 sec*) = **$31**

(5) Thirty-one dollars seems a reasonable amount for keeping a feather in the air for 50 seconds.

8. (1) How much prize money was left?

(2) $5,020,000 total prize money; Mears won $804,853; second place $335,103; third place $228,403.

(3) add and subtract

(4) $5,020,000
− 1,368,359 (*total for 1st, 2nd, and 3rd*
$3,651,641 *prizes*)

(5) It is logical that the amount of money remaining is less than the total amount.

Predicting the Question
pages 32–33

1. pay for the 5 pounds of coffee

2. How many more people lived in Homeville in 1980 than in 1905? *or* By how many people had Homeville grown?

3. c

4. a

5. b

6. Answers will vary. Sample answer: Central California is how many degrees cooler than southern California?

Seeing the Question
pages 34–35

1. 3, 2, 1 Michele used 2 pieces of wood to make a frame. One piece was 27 inches long, and the other was 20 inches long. What was the total length of the wood she used?

2. 1, 3, 2 Jeannette drove across the country for 5 days straight. She covered 400 miles the first day, 375 miles the next, then 390 miles, 560 miles, and finally 580 miles. What was Jeannette's average traveling distance per day?

3. 2, 3, 1 Shipments of cassette tapes come to the Carousel Music Store in boxes of 50. On Monday the store received 13 boxes. How many cassette tapes did the store receive that day?

4. 3, 1, 2 A man traveled 150 miles by car. He drove for 3 hours without stopping. What was his speed?

5. 3, 4, 2, 1 Mr. Johnson found that one tire from a load on his truck weighed 25 pounds. He knew that all the tires weighed the same amount. The tires weighed a total of 12,150 pounds. How many tires were in the load?

6. 2, 4, 1, 3 The teachers at Eagle Prep School are well educated. For each, the average number of years of college education is 5. Each year of college costs about $5,600. What is the amount spent on education by the typical teacher at Eagle Prep?

7. 1, 4, 3, 2 Georgiana bought a piece of material 22 feet long. She used 6 feet to make slacks. She then made dresses out of 12 feet of the material. How much extra material did she buy?

Labeling Information
pages 36–37

1. dollars

2. hours

3. ounces

4. gallons

5. Information: 183 <u>students</u>
98 <u>students</u>

Solution: 183
− 98
85 (students)

6. Information: 410 <u>pages</u>
38 <u>cents</u> or $.38

Solution: 410
× .38
3280
1,230
$155.80

7. Information: 720 <u>miles</u>
12 <u>hours</u>

Solution: **60** (mph)
12)720
72
0
0

8. Information: 31 <u>pounds</u>
57 <u>pounds</u>

Solution: 57
− 31
26 (pounds)

Sorting Out Information in Word Problems
pages 38–39

1. a

2. b

3. a

4. d

5.
$$
\begin{array}{r}
\$390.00 \\
-\ \ 80.00 \\
\hline
\$310.00
\end{array}
$$

6.
$$
\begin{array}{r}
26 \\
\times\ .30 \\
\hline
\$7.80
\end{array}
$$

7.
$$
\begin{array}{r}
464 \\
54 \\
+\ \ 28 \\
\hline
546 \text{ people}
\end{array}
$$

8.
$$
\begin{array}{cc}
12 & 21 \\
+\ 9 & \times 21 \\
\hline
21 & 21 \\
& 42 \\
\hline
& 441 \text{ students}
\end{array}
$$

Recognizing Incomplete Information
pages 40–41

1. yes

2. no the distance Thomas drives to work

3. no the number of months she had to make payments

4. no the number of guests at the party

5. yes

6. no the cost of curtain fabric

Recognizing Incomplete Information in Problems
pages 42–43

1. b $20

$$
\begin{array}{cccc}
\$3 & \$2 & \$1 & \$6 \\
\times 2 & \times 4 & \times 6 & 8 \\
\hline
\$6 & \$8 & \$6 & +\ 6 \\
& & & \hline \\
& & & \$20
\end{array}
$$

2. c

3. b 900 people

$$
\begin{array}{cc}
30 & 450 \\
\times 15 & \times\ \ 2 \\
\hline
150 & 900 \text{ people} \\
30 & \\
\hline
450 &
\end{array}
$$

4. d

5. d

Identifying the Operation
page 44

1. add **4.** divide **7.** subtract

2. subtract **5.** multiply **8.** subtract

3. multiply **6.** divide

Rounding Numbers to Choose Operations
page 45

1. 137 dictionaries — $65 rounded becomes $70; $9,570 rounded becomes $9,600, so $9,600 ÷ $70 = 137 dictionaries

2. 2,000 cows — 1,526 rounded becomes 1,500; 498 rounded becomes 500, so 1,500 + 500 = 2,000.

3. 56,000 appliances — 2,786 rounded becomes 2,800; 22 rounded becomes 20, so 2,800 × 20 = 56,000.

4. 300 yards — 1,652 rounded becomes 1,700; 1,397 rounded becomes 1,400, so 1,700 − 1,400 = 300.

5. 4,000 yards — 76 rounded becomes 80; 53 rounded becomes 50, so 80 × 50 = 4,000 yards.

Comparing Problems to Choose the Operation
pages 46–47

1. a. subtraction
 b. subtraction
 c. addition
 a; *b*; subtracting to find a difference

2. a. multiplication
 b. multiplication
 c. division
 a; *b*; multiplying to find a total

3. a. division
 b. multiplication
 c. division
 a; *c*; dividing to find an even amount

Estimating Answers to Word Problems
pages 48–49

1.
$$
\begin{array}{cc}
\textbf{Estimate} & \textbf{Real} \\
150 & 147 \\
\times\ \$.10 & \times\ \$.13 \\
\hline
\$15.00 & 441 \\
& 147 \\
\hline
& \$19.11
\end{array}
$$

2.

Estimate	Real
25	26
20) 500	18) 468
40	36
100	108
100	108

3.

Estimate	Real
$900	$871
− 200	− 199
$700	$672

4.

Estimate	Real
$1,300	$1,320
200	229
1,900	1,898
+ 2,100	+ 2,102
$5,500	$5,549

Deciding Whether the Answer Makes Sense
pages 50–51

1. c or d   **3.** b   **5.** b

2. a or b   **4.** b or c

CHAPTER 4: MEASUREMENTS

Comparing Measurements
pages 52–53

1. a   **6.** a   **11.** ton

2. c   **7.** b   **12.** pint

3. b   **8.** c   **13.** gallon

4. b   **9.** yard   **14.** kilometer

5. c   **10.** century   **15.** tablespoon

Converting Measurements
pages 54–55

1. a. 6,000 pounds   2,000
   b. multiply   × 3
                      6,000

2. a. 180 seconds   60
   b. multiply   × 3
                      180

3. a. 2 pounds   2
   b. divide   16) 32
                     32

4. a. 5 feet   5
   b. divide   12) 60
                     60

5. a. 52 weeks   52
   b. divide   7) 364
                     35
                     14
                     14

6. a. 80 years   10
   b. multiply   × 8
                      80

7. a. 2 days   2
   b. divide   24) 48
                     48

8. a. 24 fluid ounces   8
   b. multiply   × 3
                      24

9. a. 15,840 feet   5,280
   b. multiply   × 3
                      15,840

10. a. 2 years   2
    b. divide   12) 24
                      24

Drawing Pictures
pages 56–57

1. 2 lengths + 2 widths = picture
   $2(12) + 2(11) =$ **46 in**

2. **(1)** 2 × 4 ft = 8 ft curtain
   (2) 3 yd × 3 ft = 9 ft
   (3) 9 ft − 8 ft = **1 ft left**

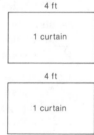

3. **(1)** 9 + 9 + 3 = framing per door
        21 ft = framing
   (2) 21 ft × 8 doors = **168 ft**

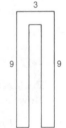

4. a. 92 ft   53 ft   53 ft
   b. 12 ft   × 92 ft   × 12 ft
   c. 53 ft   106   106
   d. 53 ft   477   53
                   4,876 sq ft   636 sq ft

   4,876 *(front yard)*
   − 636 *(driveway)*
   4,240 sq ft

Getting Information from Charts and Tables
pages 58–59

1. **$2.50** (1 lb 7 oz is between 1 and 2 lb)
2. **$5.00** (20 oz is between 1 and 2 lb $2.50 + $2.50 = $5.00)
3. **$5.00** (16 + 13 + 4 = 33 oz; between 2 and 5 lb)
4. **$1.00** (4 + 2 + 1 = 7 oz)
5. **$7.50** (2 lb 10 oz + 3 lb 6 oz = 5 lb 16 oz or 6 lb)
6. petite, medium
7. regular, x-large
8. 38, 30, 41
9. petite, 8

Multistep Measurement Problems
pages 60–61

1. Problem 1

$$\begin{array}{r} \underline{\ 50}\ \text{calories in a half-pint of strawberries} \\ 2)\overline{100}\ \text{calories in 1 pint of strawberries} \end{array}$$

Problem 2	Problem 3
150 whole milk	2,500
80 fish	− 545
50 butter	**1,955 calories**
215 cottage cheese	
+ 50 strawberries	
545 calories before dinner	

2. Problem 1

$$\begin{array}{r} \underline{\ 50}\ \text{calories in a half-pint of strawberries} \\ 2)\overline{100}\ \text{calories in 1 pint of strawberries} \end{array}$$

Problem 2	Problem 3
Clare:	Jodi:
2 diet pop	80 fish
215 cottage cheese	50 butter
+ 50 strawberries	+ 120 lowfat milk
267 calories	**250 calories**

Problem 4: Clare had more calories.

3. **Problem 1 Question:**
How many calories did Robyn consume that day?

Problem 2 Question:
How far from her allowance was she?

Problem 1	Problem 2
0 coffee	1,200
120 yogurt	− 723
215 cottage cheese	**477 calories**
50 strawberries	
1 diet pop	
120 fish	
215 cottage cheese	
+ 2 diet pop	
723 calories	

Using Formulas
pages 62–63

1. (1) **Question:** How much litter does George need to buy?
   (2) **Formula:** Since I need to find the inside space of the litter box, I choose the volume formula.
   $V = l \times w \times h$
   (3) **Necessary Information:** 20 inches wide; 32 inches long; 4 inches high; halfway
   (4) **Solution:** $V = 32 \times 20 \times 4 = 2,560$ cu in
   $\frac{1}{2}V = \frac{1}{2} \times 2,560 = $ **1,280 cu in**
   (5) **Check For Logic:** 1,280 cubic inches of litter seems reasonable to fill a litter box halfway with dimensions of $32 \times 20 \times 4$ inches.

2. (1) **Question:** How much lace does Lilli Mae need to buy?
   (2) **Formula:** Since I need to find the distance around the tablecloth, I choose the perimeter formula.
   $P = 2l + 2w$
   (3) **Necessary Information:** 2 yards wide; 3 yards long
   (4) **Solution:** $P = 2(3) + 2(2) = 10$ yd
   (5) **Check for Logic:** 10 yards seems a reasonable length for attaching lace around a tablecloth.

3. (1) **Question:** How large should Victor's piece of glass be?
   (2) **Formula:** Since I need to find the surface space, I choose the area formula.
   $A = l \times w$
   (3) **Necessary Information:** 8 inches wide by 13 inches long
   (4) **Solution:** $A = 13 \times 8 = 104$ sq in
   (5) **Check for Logic:** 104 square inches seems reasonable for a piece of glass to cover Victor's picture.

4. (1) **Question:** How many containers of 4-pack golf balls will fit in a full box?
   (2) **Formula:** Since I need to find the inside space of a box, I choose the volume formula. $V = l \times w \times h$
   (3) **Necessary Information:** 4 inches long, 2 inches wide, 2 inches high; 11 inches high, 12 inches wide, 16 inches long
   (4) **Solution:** V (of containers) $= 4 \times 2 \times 2 = 16$ cu in
   V (of boxes) $= 16 \times 12 \times 11 = 2,112$ cu in
   $$\begin{array}{r} \textbf{132 containers} \\ 16)\overline{2,112} \end{array}$$
   (5) **Check for Logic:** 132 containers seems a reasonable number to fill a box that measures 2,112 cubic inches.

CHAPTER 5: DECIMALS

Knowing When to Estimate
page 67

1. **estimate** Large profits generally are expressed in round figures, not in exact dollar amounts.

2. **estimate** College costs change from year to year, so an exact figure could not be accurate.

3. **exact** Mechanical parts must be precise for machines to operate.

4. **estimate** A budget is usually based on estimated figures.

5. **exact** Precipitation is usually measured in tenths of an inch; to show whether a record has been broken, measurements must be exact.

6. **exact** Buttonholes must be the exact size for a button to fit.

7. **estimate** The number of plants (a dozen or so) is an estimate, so an exact figure is not necessary.

8. **exact** Judging an Olympic athlete's performance requires split-second precision.

Estimating Answers
pages 68–69

1. QUESTION: What is Mort's new balance?
NECESSARY INFORMATION:
$170, $40, and $30
NUMBER SENTENCE: $170 − ($40 + $30) = ?
ESTIMATE: $170 − $70 = **$100 estimate**
CHECK:
$171.19 − ($39.88 × $26.14) = **$105.17 exact**

2. QUESTION: How long was the piece of metal?
NECESSARY INFORMATION:
.3 in, .4 in, .7 in, 2.5 in
NUMBER SENTENCE:
2.5 in − (.3 + .4 + .7) = ?
ESTIMATE: 2.5 − 1.4 = **1.1 in**
CHECK:
2.50 − (.290 + .413 + .651) = **1.146 in**

3. QUESTION: How much money did the butcher make on pork chops?
NECESSARY INFORMATION: 27 lb, $3/lb
NUMBER SENTENCE: 27 lb × $3 = ?
ESTIMATE: 27 lb × $3 = **$81**
CHECK: 27.28 lb × $2.79 lb = **$76.11**

4. QUESTION: How many pieces can Ellie cut from the roll?
NECESSARY INFORMATION: 38 in, 760 in
NUMBER SENTENCE: 760 in ÷ 38 in = ?
ESTIMATE: 760 in ÷ 38 in = **20 pieces**
CHECK:
759.5 in ÷ 37.5 in = **20.25 or 20 pieces**

5. QUESTION: What was Benjy's average?
NECESSARY INFORMATION: .300, .280, .250
NUMBER SENTENCE:
(.300 + .280 + .250) ÷ 3 = ?
ESTIMATE: (.300 + .280 + .250) ÷ 3 = **.277**
CHECK: (.297 + .278 + .248) ÷ 3 = **.274**

Comparing Decimal Numbers
page 70

1. $.04
2. $.03
3. $1.78
4. $12.00
5. 7.70 yd is larger
6. 16.90 ft is larger.
7. .300 in is largest.
8. .660 oz is largest.

Metric Measurement with Decimals
page 71

1. kilometer
2. milligram
3. 10 deciliters make 1 liter.
4. 100 centimeters make 1 meter.
5. 1,000 milligrams make 1 gram.
6. **a** 250 milliliters
7. **b** 25 meters
8. **b** 140 milligrams

Paychecks and Checking Accounts
pages 72–73

1. Gross pay is the pay that Lisa receives **before** taxes and social security are taken out.

2. Net pay is the pay that Lisa receives **after** taxes and social security are taken out.

3. Net pay is found by **subtracting** the deductions from the gross pay.

4. The deductions from Lisa's pay are **federal income tax, state income tax**, and **social security**.

5. **a.** division
   b. $8,064.00 ÷ $201.60 = **40 weeks**

6. **a.** multiplication and addition.
   b. ($.15 × 40) + $201.60 = **$207.60**

7. **a.** addition
   b. $25.87 + $4.29 + $16.08 = **$46.24**

8. $577.56 + $201.60 = **$779.16**

9. $779.16 − $12.00 = **$767.16**

10. $767.16 − $100.00 = **$667.16**

11. $667.16 − $30.47 = **$636.69**

12. $636.69 + ($201.60 − $50) = **$788.29**

13. $788.29 − $4.25 = **$784.04**

Finding Total Cost
page 74

1. $4.95 × 3 = **$14.85**

2. $.35 × 5 = **$1.75**

3. $1.09 × 16 = **$17.44**

4. $6.45 × 7 = **$45.15**

5. $2.79 × 3.2 = **$8.93**

6. **a $2.00**
   ESTIMATE: 5 lb × $.40 = **a. $2.00**

7. **c 40 cents**
   ESTIMATE: $4.00 × .10 = **c. $.40**

8. **b $.60**
   ESTIMATE: $.30 × 2 = **b. $.60**

Finding the Best Buy
page 75

1. **a.** $2.25 ÷ 12 = $.19
   b. $2.99 ÷ 24 = $.12
   c. $3.69 ÷ 36 = $.10
   So **c** is the best buy.

2. **a.** $1.10 ÷ 50 = $.022
   b. $.88 ÷ 75 = .011 = $.01
   c. $3.50 ÷ 150 = .023 = $.02
   So **b** is the best buy.

3. **a.** $.70 ÷ 6 = $.12
   b. $1.49 ÷ 10 = $.15
   c. $1.59 ÷ 12 = $.13
   So **a** is the best buy.

4. **a.** $2.39 ÷ 60 = .039 = $.04
   b. $2.99 ÷ 100 = .029 = $.03
   c. $6.00 ÷ 150 = $.04
   So **b** is the best buy.

Multistep Problems with Decimals
pages 76–77

1. Problem 1: What is the total monthly cost of the braces? $180 + $100 = $280

   Problem 2: What is the cost over 24 months? $280 × 24 = **$6,720**

2. Problem 1: How many 8-ounce cans make 48 ounces of punch? 48 ÷ 8 = 6

   Problem 2: How many cans of each of the three juices will make the punch?
   6 ÷ 3 = **2 cans each**

3. Problem 1: How many 4-ounce servings are in a 24-oz box?
   24 oz ÷ 4 oz = 6 servings

   Problem 2: How many calories does a 24-ounce box contain?
   6 × 110 = **660 calories**

   The amounts of protein and potassium are not necessary because they do not measure calories.

4. Problem 1: How much do the Bonzers currently pay to rent 50 movies? 50 × $2 = $100

   Problem 2: How much would the Bonzers pay to rent 50 movies if they joined the Video Club? $50 × 1 = $50

   Problem 3: How much would the Bonzers pay in total if they joined the Video Club?
   $50 + $25 = $75

   Problem 4: How much will they save if they join the club? $100 − $75 = **$25**

CHAPTER 6: FRACTIONS

Writing Fractions
pages 78–79

1. 3 dimes = $\frac{3}{10}$ of a dollar

2. 2 pennies = $\frac{2}{100}$ or $\frac{1}{50}$ of a dollar

3. 3 quarters = $\frac{75}{100}$ or $\frac{3}{4}$ of a dollar

4. 1 half-dollar = $\frac{50}{100}$ or $\frac{1}{2}$ of a dollar

5. 4 nickels = $\frac{20}{100}$ or $\frac{4}{20}$ or $\frac{1}{5}$ of a dollar

6. $\frac{37}{50}$

7. $\frac{8}{9}$

8. $\frac{2}{3}$

9. $\frac{4}{15}$

10. $\frac{121}{330}$

Adding and Subtracting with Common Denominators
pages 80–81

1. 24

2. 24

3. 21

4. 72

5. 21

6. 20

7. $\frac{3}{7} + \frac{4}{21} = \frac{9}{21} + \frac{4}{21} = \frac{13}{21}$

8. $\frac{2}{3} + \frac{5}{13} = \frac{26}{39} + \frac{15}{39} = \frac{41}{39} = 1\frac{2}{39}$

9. $\frac{5}{8} - \frac{1}{2} = \frac{5}{8} - \frac{4}{8} = \frac{1}{8}$

10. $\frac{4}{5} - \frac{1}{20} = \frac{16}{20} - \frac{1}{20} = \frac{15}{20}$ or $\frac{3}{4}$

Multiplying Fractions
pages 82–83

1. EXPRESSION:

$\frac{5}{6}$ of 204 minutes = time spent

SETUP: $\frac{5}{6} \times 204 = \boxed{?}$

SOLUTION: $\frac{5}{6} \times 204 = $ **170 minutes**

2. EXPRESSION: $\frac{2}{5}$ of 550,000 cars = cars needing repairs

SETUP: $\frac{2}{5} \times 550{,}000 = \boxed{?}$

SOLUTION: $\frac{2}{5} \times 550{,}000 = $ **220,000**

3. EXPRESSION:

$\frac{3}{4}$ of $5\frac{1}{3}$ yd. of lace = amount bought

SETUP: $\frac{3}{4} \times 5\frac{1}{3}$ yd $= \boxed{?}$

SOLUTION: $\frac{3}{4} \times \frac{16}{3} = $ **4 yd**

4. EXPRESSION:

$\frac{5}{8}$ of $1\frac{1}{5}$ lb chocolates = chocolates eaten

SETUP: $\frac{5}{8} \times 1\frac{1}{5} = \boxed{?}$

SOLUTION: $\frac{5}{8} \times \frac{6}{5} = \frac{3}{4}$ **lb**

Dividing Fractions
pages 84–85

1. EXPRESSION:

$\frac{2}{3}$ of total force = 48 employees

SETUP: $\frac{2}{3}$ of $\boxed{?} = 48$

$\boxed{?} = 48 \div \frac{2}{3} = 48 \times \frac{3}{2}$

SOLUTION: $48 \times \frac{3}{2} = $ **72 employees**

2. EXPRESSION:

$\frac{3}{5}$ of combined income = \$13,440

SETUP: $\frac{3}{5}$ of $\boxed{?} = \$13{,}440$

$\boxed{?} = \$13{,}440 \div \frac{3}{5} = \$13{,}440 \times \frac{5}{3}$

SOLUTION: $\$13{,}440 \times \frac{5}{3} = $ **\$22,400**

3. EXPRESSION:

$\frac{7}{8}$ of a pound of bananas = \$.56

SETUP: $\frac{7}{8}$ of $\boxed{?} = \$.56$

$\boxed{?} = \$.56 \div \frac{7}{8} = \$.56 \times \frac{8}{7}$

SOLUTION: $\$.56 \times \frac{8}{7} = $ **\$.64**

4. EXPRESSION:

$\frac{7}{8}$ of the calories = 105 calories

SETUP: $\frac{7}{8}$ of $\boxed{?} = 105$

$\boxed{?} = 105 \div \frac{7}{8} = 105 \times \frac{8}{7}$

SOLUTION: $105 \times \frac{8}{7} = $ **120 calories**

Deciding Whether to Multiply or Divide
pages 86–87

1. EXPRESSION:

$\frac{2}{3}$ of $12\frac{1}{2}$ cups = amount of flour used

OPERATION: multiply

REASON: part not given

SOLUTION: $\frac{2}{3} \times 12\frac{1}{2} =$
$\frac{2}{3} \times \frac{25}{2} = \frac{25}{3} = 8\frac{1}{3}$ **cups used**

2. EXPRESSION:

$\frac{4}{5}$ of a tank of gas = 18.4 gallons

OPERATION: divide

REASON: whole not given

SOLUTION: $\frac{4}{5}$ of $\boxed{?} = 18.4$

$\boxed{?} = 18.4 \div \frac{4}{5} =$
$18.4 \times \frac{5}{4} = $ **23 gallons**

3. EXPRESSION:

$\frac{5}{10}$ of questions = 25 questions

OPERATION: divide
REASON: whole not given

SOLUTION: $\frac{5}{10}$ of ? = 25

$? = 25 \div \frac{5}{10} =$

$25 \times \frac{10}{5} =$ **50 questions**

4. EXPRESSION:

$\frac{2}{7}$ of $357.35 = Bruskas' down payment

OPERATION: multiply
REASON: part not given

SOLUTION: $\frac{2}{7} \times \$357.35 =$ **\$102.10**

Ratios and Proportions
page 88

1. $\frac{7}{8}$, 7:8

2. $\frac{9}{10}$, 9:10

3. $\frac{10}{100}$, 10:100

4. $\frac{1}{3}$, 1:3

5. ⌐

6.

7. **c** sale price

8. **b** 1

9. **b** ounces

Setting Up Proportions
page 89

1. $\frac{\$144}{12\text{ hr}} = \frac{x}{40}$ or $\frac{12}{144} = \frac{40\text{ hr}}{x}$

2. $\frac{10\text{ teachers}}{180\text{ students}} = \frac{x}{216}$ or $\frac{180}{10} = \frac{216\text{ students}}{x}$

3. $\frac{30\text{ problems}}{10\text{ min}} = \frac{21}{x}$ or $\frac{10}{30} = \frac{x}{21}$

4. $\frac{3\text{ lb}}{\$6.69} = \frac{2.5\text{ lb}}{x}$ or $\frac{\$6.69}{3\text{ lb}} = \frac{x}{2.5\text{ lb}}$

5. $\frac{10}{8} = \frac{15}{x}$ or $\frac{8}{10} = \frac{x}{15}$

Solving Proportions
page 90

1. **c** $x = 21$ 　　$\frac{1}{7} = \frac{3}{x}$
$x = 21$

2. **c** $x = 60$ women 　$\frac{10}{40} = \frac{15}{x}$
$10x = 600$
$x = 600 \div 10$
$x = 60$

3. **b** 16 birds 　$\frac{12}{3} = \frac{x}{4}$
$3x = 48$
$x = 48 \div 3$
$x = 16$

4. **d** 45 lb 　$\frac{10}{15} = \frac{30}{x}$
$10x = 450$
$x = 450 \div 10$
$x = 45$ lb

Solving Word Problems Using Proportions
page 91

1. $\frac{2\text{ files}}{10\text{ hours}} = \frac{25\text{ files}}{x\text{ hours}}$
$2x = 250$
$x = 250 \div 2 =$ **125 hours**

2. $\frac{12\text{ feet rope}}{1\text{ foot post}} = \frac{x\text{ feet rope}}{7\text{-foot post}}$
$x =$ **84 feet of rope**

3. $\frac{13\text{ students}}{39\text{ books}} = \frac{156\text{ students}}{x\text{ books}}$
$13x = 6084$
$x = 6084 \div 13 =$ **468 books**

4. $\frac{3\text{ cups sugar}}{2\text{ cups water}} = \frac{9\text{ cups sugar}}{x\text{ cups water}}$
$3x = 18$
$x = 18 \div 3 =$ **6 cups water**

CHAPTER 7: PERCENTS

When to Use Decimals, Fractions, and Percents
page 92

1. d 　　　　4. d

2. p 　　　　5. f

3. d 　　　　6. p

Percents as Ratios and Fractions
page 93

1. 72 out of every 100 parts; $\frac{72}{100}$

2. 9 out of every 100 people; $\frac{9}{100}$

3. 17 out of every 100 scores; $\frac{17}{100}$

4. 98 out of every 100 students; $\frac{98}{100}$

5. $33\frac{1}{3}$ out of every 100 houses; $\frac{33\frac{1}{3}}{100}$

6. 6.5 out of every 100 cents; $\frac{6.5}{100}$

7. $300 for every $100 or $3.00 for every $1.00; $\frac{300}{100}$

Percents As Decimals
page 94

1. .01
2. $.25
3. .08
4. .129
5. 1.10
6. 3
7. 7.6%
8. 10%
9. 200%
10. 31.1%

Using the "Of" Formula with Percents
pages 95–97

1. EXPRESSION:
   6.5% of the purchase = sales tax
   SOLUTION: .065 × $132.76 = **$8.63**

2. EXPRESSION:
   40% of registered voters = cast ballots
   SOLUTION: .40 × 3,905 = **1,562 voters**

3. EXPRESSION: 75% of 300 questions = the number required to earn a C
   SOLUTION: .75 × 300 = **225 questions**

4. EXPRESSION: 12% of last year's amount of snow = 10.08 inches
   SOLUTION: .12 of ? = 10.08
   ? = 10.08 ÷ .12 = **84 in**

5. EXPRESSION: 30% of the price of the dinette set = $300
   SOLUTION: .30 of ? = $300
   ? = $300 ÷ .30 = **$1,000**

6. EXPRESSION: 150% of a home's former price = $120,000
   SOLUTION: 1.50 of ? = $120,000
   ? = $120,000 ÷ 1.50 = **$80,000**

7. EXPRESSION: What percent of $585 = $100
   SOLUTION: $100 ÷ $585 = **17%**

8. EXPRESSION: $7 off on $21 shirt = what percent of savings
   SOLUTION: $7 ÷ $21 = **$33\frac{1}{3}$%**

Using the Percent Circle
pages 98–99

1. 4,000 ÷ 8,000 = **50%**

2. $19,000 × .085 = **$1,615**

3. 1,100 stalks × .15 = **165 stalks**

4. $107.10 ÷ .07 = **$1,530**

5. $182 ÷ 3.50 = **$52**

6. 475 mi ÷ .25 = **1,900 mi**

7. $6.75 ÷ $125.00 = **5.4%**

Percent of Increase or Decrease
pages 100–101

1. $.30 − $.20 = $.10
   $.10 ÷ $.20 = **50%**

2. $7 − $6 = $1
   $1 ÷ $6 = **16.6% or $16\frac{2}{3}$%**

3. $90,000 − $50,000 = $40,000
   $40,000 ÷ $50,000 = **80%**

4. $120 − $40 = $80
   $80 ÷ $40 = **200%**

5. growth, raise, profit, markup

6. $10,000 − $8,000 = $2,000
   $2,000 ÷ $10,000 = **20%**

7. $550 − $495 = $55
   $55 ÷ $550 = **10%**

8. 12,000 − 3,000 = 9,000
   9,000 ÷ 12,000 = **75%**

9. 8 − 7 = 1
   1 ÷ 8 = **$12\frac{1}{2}$%**

10. depreciation, discount, lost, shrinkage

Original and Sale Price
pages 102–103

1. 100% − 25% = 75%
   $60 ÷ .75 = **$80**

2. 100% − 5% = 95%
   $10.45 ÷ .95 = **$11.00**

3. 100% − 30% = 70%
   $24 ÷ .70 = **$34.29**

4. 100% − 75% = 25%
   $200 ÷ .25 = **$800**

5. $12,550 × .20 = $2,510
   $12,550 − $2,510 = **$10,040**

6. $105 × .40 = $42
   $105 − $42 = **$63**

7. $120 × .125 = $15
$120 − $15 = **$105**

8. $50 × .30 = $15
$50 − $15 = **$35**

The Interest Formula
pages 104–105

1. $1,000.00 × .0575 × 3 = **$172.50**

2. $500.00 × .065 × 1 = **$32.50**

3. $1,800 × .05 × 2.5 = **$225**

4. $835.00 × .075 × 1 = **$62.63**

5. $172.50; $1,000.00; **$1,172.50**

6. $32.50; $500.00; **$532.50**

7. $225; $1,800; **$2,025**

8. $62.63; $835.00; **$897.63**